Energy for Planet Earth

Energy for Planet Earth

· · ·

READINGS FROM
SCIENTIFIC AMERICAN MAGAZINE

W. H. FREEMAN AND COMPANY
New York

Cover image by George V. Kelvin

Library of Congress Cataloging-in-Publication Data

Energy for planet earth.
    p.   cm.
    Collection of articles from Scientific American.
    Includes bibliographical references and index.
    ISBN 0-7167-2235-6
    1. Power resources. 2. Renewable energy
sources.   I. Scientific American.
TJ163.24.E539   1991
333.79—dc20   90-23226
                                    CIP

The eleven chapters and the epilogue in this book
originally appeared as articles and the closing essay in the
September 1990 issue of SCIENTIFIC AMERICAN.

Printed in the United States of America

1 2 3 4 5 6 7 8 9 0 RRD 9 9 8 7 6 5 4 3 2 1

CONTENTS

Foreword

If you were told that the aircraft that you were boarding had a 10 percent chance of crashing, you would probably choose another way to reach your destination. Yet those in government, industry, and academia who question the need to respond to the dangers of global warming do so by choosing to focus on the uncertainties in the climatic models. They question the value of subjecting the world to major economic and political dislocations in an effort to alter predictions based on a model with an admittedly large margin of error. They are hoping that the plane won't crash.

However, once it begins, global warming is irreversible. The risk of a future shaped by migrating climate zones, rising seas, collapsing economies and political turbulence among countries, large and small who are armed with nuclear weapons, is too great a risk to take.

Technologically, there are many options. Thanks to bottom-line pressures and improved control systems, industrial energy use has actually declined per unit of output. Factories and offices have begun to be replaced by smart buildings whose comfortable interior climates are maintained by computer systems. Advances in materials, engines, traffic control and the regional organization of many transit systems promises to ameliorate the impact of the world's fleet of 500 million automobiles.

The traditional fuel stock can now feed advanced heat generators that yield maximum amounts of energy. Such devices reduce the amount of fossil fuel consumed, buying time for other, more environmentally kind, technologies to mature. A future without nuclear energy would be a difficult one to engineer, but unless a new generation of ultrasafe reactors comes on line and solutions are found for the problems of waste disposal and plant security, it may be a prospect. As the price of energy begins to reflect the cost of environmental degradation, power from renewable resources is increasingly economically, as well as environmentally, attractive. Windmill farms, solar panels, and alcohol feedstocks make solar energy available to us, either directly or indirectly, for a variety of uses.

Can human societies mobilize their economic and political strengths to implement such solutions? In the market economy of the northern hemisphere, economic and political conservatism often checks a rising awareness of the need to take action. Yet, despite this conservatism, both business and government have been implementing tougher emissions standards for automobiles, as well as for the release of carbon dioxide, chlorofluorocarbons and other greenhouse gases.

In the southern hemisphere, the burgeoning populations of developing countries need to vault over the traditional phase of economic development, in which vast quantities of fossil fuels are consumed in order to provide a higher standard of living. One answer consists of intensive, well-planned use of locally available and implementable energy sources, instead of traditionally massive, grandiose power systems.

A different challenge confronts the citizens of the diverse emerging democracies that stretch from the Brandenberg Gate to the coast of China. There the production and consumption of energy have been carried out by planners oblivious to the environmental consequences. The challenge is to undo the damage in the context of economic, political and legal systems that are currently in a state of flux.

How can we best supply energy for all of planet Earth? The problem is not a technological one; it is political and moral. We must match solutions to their various cultural and social contexts. It must be universally recognized that the relationship between human society and energy has entered a profound transformation. Our resources are no longer infinite, and neither is the capacity of the environment to absorb the consequences of exploitation.

Our ability to meet the challenges posed by these circumstances is immense, and the timing is critical. As East-West tensions wane, regional conflicts suppressed by the superpower agendas have flared to life. But what could more powerfully unify our species than global cooperation to produce an energy economy in the context of a sustainable environment that benefits all nations?

Jonathan Piel
For the Board of Editors

Energy for Planet Earth

Energy for Planet Earth

*Our ability to meet the world's energy needs
without destroying the planet on which we live is examined.
With the right incentives, much can be done.*

. . .

Ged R. Davis

As early as 400,000 B.C., fire was kindled in the caves of Peking Man. Revered as a deity and the basis of many myths, fire has been an essential element in the technologies on which civilized societies are founded. Engines driven by fossil fuels have replaced human and animal muscle, precipitating the rise of industrialized societies. Today cities, industrial facilities and transportation networks could not function without regular supplies of energy.

As long as the number of human beings remained small and energy needs were limited to cooking and heating, energy could be exploited without serious disruption to the atmosphere, hydrosphere and geosphere. Now amplified by a growing population, energy use has become a potentially destructive force, locally because emissions contaminate air, water and soil and globally because there is the possibility that energy use may enhance the greenhouse effect. We face a dilemma: properly used, energy technologies serve as instruments for realizing material well-being across the planet, but continuation of current trends could lead to a degraded environment, yielding a mean and uncertain existence.

Coming to grips with the challenge of ensuring an adequate, safe energy supply is a theme that pervades this book. Environmental concerns are not new, but our understanding of the planet has changed by virtue of our newfound ability to measure ever smaller concentrations of substances and to assess their implications—for human beings and for the earth as a whole. And we have come to realize that population growth and its rising demands may transform the planet in ways comparable to the effects of long-term geologic forces.

In order to understand the magnitude of the challenge, it helps to understand where energy comes from and the purpose that it serves in our lives. Almost all available energy can be traced either to the sun (see Figure 1.1)—fossil fuels, biomass, wind and incoming radiation—or to the processes of cosmic evolution preceding the origin of the solar

Figure 1.1 SUN IS THE SOURCE from which almost all energy on the earth is derived. It is the driving force behind photosynthesis, which converts the sun's radiant energy to chemical energy, making plant and ultimately all animal life possible. Photosynthesis is also responsible for the formation of fossil fuels. More directly, the sun provides energy that can be captured in the form of solar power, hydropower and wind power.

system—nuclear power. Smaller, less significant amounts are derived from lunar motion (tidal power) and from the earth's core (geothermal power).

If society could exploit only a small portion of the solar radiation that strikes the earth's surface every year, which is equivalent to 178,000 terawatt-years (or about 15,000 times the world's present energy supply), our energy problems would be solved. Of that amount, however, 30 percent is reflected back to space and 50 percent is absorbed, converted to heat and reradiated (see Figure 1.2). The 20 percent that remains powers the hydrologic cycle. Only a very small share (.06 percent) of solar radiation powers photosynthesis, from which all life and fossil fuels are ultimately derived. Currently renewables (including hydropower and biomass) account for 18 percent of the world's energy needs and nuclear power 4 percent; the remainder is met by fossil fuels.

Few people care about the source of our energy supply except when it is disrupted, but virtually all of us care about energy services, which range from the basic needs demanded by human beings everywhere—cooking, heating and lighting—to the hallmarks of modern society—motors, appliances, wide-ranging mobility and various industrial processes. Because the world cannot function without regular supplies of energy, a significant section of the global economy is devoted to providing these services when and where required.

Lighting a room, for instance, is not achieved merely by flicking a switch; it is the last step in a long chain of conversion events. Energy resources —for example, the unrefined oil and natural gas recovered from wells driven deep into the earth's crust and the coal that is sandwiched between terrestrial sediments—must first be extracted (see Figure 1.3). The primary energy (crude oil, say) is then transported to a refinery to be processed into a wide range of products, and from there fuel oil is shipped to a power plant to be burned (and thus converted from chemical to thermal energy). The heat produced during combustion powers a turbine, which in turn drives an electric generator (converting thermal to mechanical to electric energy). Eventually the electricity travels through wires until it reaches the end-use appliance—the incandescent lamp— where it is transformed into radiant energy.

The uneven distribution of the world's fossil fuels (oil, natural gas and coal) necessitates a flourishing worldwide trade in energy commodities; some 44 percent of oil, 14 percent of gas and 11 percent of coal consumed are traded internationally. Extensive distribution systems exist to service this trade and ensure that resources reach the consumer. Natural gas is transported over land through some one million kilometers of trunk pipelines and oil through 400,000 kilometers of pipes, excluding local distribution systems. About 2,600 tankers ply the world's oceans carrying crude oil; another 65 vessels deliver liquid natural gas around the world.

As a result of such global demand, fossil fuels are being depleted at a rate that is 100,000 times faster than they are being formed. Coal's share of the world energy supply has already peaked; in 1920 it accounted for more than 70 percent of fuel use, but today it meets only 26 percent of global energy needs (see Figure 1.4). Oil peaked in the early 1970's at slightly more than 40 percent (today it is 38 percent). The portion currently allotted to natural gas (19 percent) is expected to rise further. Although the remaining amount of recoverable fossil fuel is thought to equal some 10 trillion barrels of oil— enough to last another 170 years at present consumption rates—the supply will eventually run out, and in the interim (if it is fully combusted) the prospect presents a possible threat to the environment.

How do we reconcile our burgeoning demand for energy with the need to maintain a viable global ecosystem? There is no solution, as yet. The uncertainties surrounding environmental problems, such as climate change, and the diversity of views on the relative trade-offs between economic growth and the environment can lead to a multiplicity of policies and projections for energy supply and use.

To simplify matters, I would like to explore two possible routes to the future. The "consensus" view, held by many, rests on a continuation of present trends, whereas the "sustainable world" view presumes that global environmental issues will be on the international agenda by the mid-1990's. Underlying both scenarios is the assumption that by 2010 world population will total seven billion and gross world product will have doubled (see Figure 1.5).

In the consensus view, consumer habits and ways of life are not expected to change significantly, and the price of crude oil will probably rise gradually, although its trajectory may be volatile. World energy consumption is expected to increase by 50 or 60 percent by 2010, and the global mix of fuels is to remain substantially the same as today. Thus, global carbon dioxide (CO_2) emissions would also increase

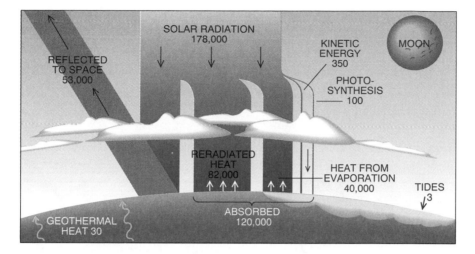

Figure 1.2 SOLAR RADIATION is equal to 178,000 tera-watts a year. Of this, 30 percent is reflected back to space; another 50 percent is absorbed, converted to heat and rera-diated. The remaining 20 percent creates wind, powers the water cycle and drives photosynthesis. Some energy, in the form of geothermal heat, can be tapped from the earth's core; a tiny amount (generated by the moon's gravitational pull) exists as tidal power. Estimates for the potential of commercial renewable resources suggest they may eventually increase from today's level of about one terawatt-year a year to 10 or perhaps 15 terawatt-years a year.

by 50 or 60 percent. Implicit in the consensus view is that "more of the same" is sustainable and that climate change is either not a serious issue or it is something to which humans can adapt.

There is much uncertainty surrounding the issue of global warming, but if studies do confirm a link between CO_2 emissions and climate change, then the consensus view of development could come at great cost. For example, a report recently issued by the Intergovernmental Panel on Climate Change concludes that "business as usual" developments could lead to a global mean temperature increase during the next century of about .3 degree Celsius each decade, with significant impact on natural and human systems. Experience tells us that public pol-icy must be not only adaptive but also anticipatory. This is the basis of the sustainable-world view.

Yet the system that enables society to produce energy does not lend itself readily to a flexible, quick response. An intractable infrastructure (power

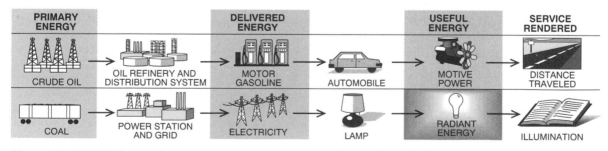
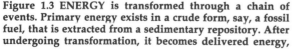

Figure 1.3 ENERGY is transformed through a chain of events. Primary energy exists in a crude form, say, a fossil fuel, that is extracted from a sedimentary repository. After undergoing transformation, it becomes delivered energy, which is made available to the consumer, who then con-verts it into useful forms and then finally into energy services, which are the desired end.

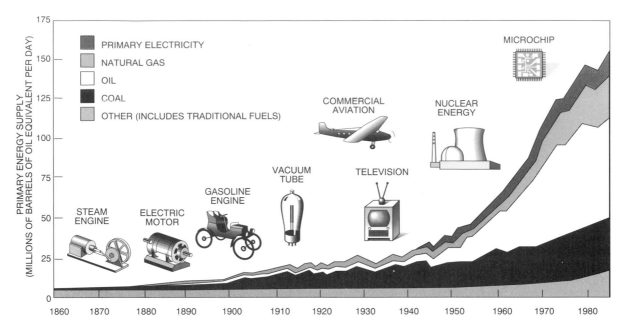

Figure 1.4 RATE OF PRIMARY ENERGY USE and the relative contributions of different sources reflect the evolution of technology as well as the growth of the human population. The rapid rise of oil after World War II, for example, is indicative of the rise of mass transportation and industry. Similarly, the growth of electricity in the late 1960's parallels the rise of a services-oriented economy. Although fossil fuels still dominate the primary energy supply, coal's share peaked around 1920, when it provided more than 70 percent of all the fuel consumed; oil's share peaked in the early 1970's at slightly more than 40 percent. Natural gas, which is less polluting than either oil or coal, is expected to contribute more to global energy use.

plants may last from 20 to 40 years), long lead times (from blueprint to operation for many energy projects may span a dozen or more years) and entrenched public perceptions (of costs, environmental acceptability and need) all make for a system laden with inertia. Projects currently under way, which originated years ago, will dominate the scene for years to come. Still, there is reason to think change is possible.

History itself is defined by rapid technological evolution. Whereas the planet was home to only a few hundred million human beings at the start of the industrial revolution, it now shelters some five billion people, who occupy around a billion dwellings, drive 500 million motor vehicles and expend much effort to produce a variety of industrial products to further their well-being. Total delivered energy (the amount that reaches the end user—the electricity, say, needed to light a lamp or the natural gas needed to heat a home) rose from the equivalent of about eight million barrels of oil a day in 1860 to 123 million barrels a day in 1985. If firewood is excluded from the calculation, delivered energy (mostly coal, oil, natural gas and electricity) has increased nearly 60-fold. And energy services have increased at a much more rapid pace than delivered energy, largely as a result of improvements in end-use efficiencies.

The demand for energy will be compounded further by the economic transition envisaged for the developing nations, where 90 percent of the world's population growth will take place. In the poorest economies the average person consumes an amount of traditional fuel (such as wood and other organic wastes, most of which are gathered rather than bought) equivalent to one or two barrels of oil a year. As countries industrialize and urbanize, commercial fuels displace traditional ones. The average person in a developing country annually uses the equivalent of one or two barrels of oil of commercial fuel (which is purchased on the open market). In contrast, the number jumps to between 10 and 30

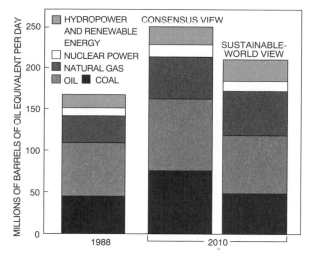

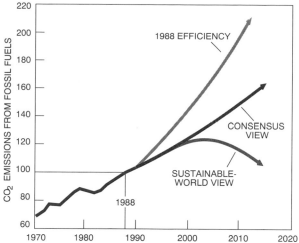

Figure 1.5 TWO SCENARIOS for the global energy mix in 2010 are shown (*left*). The consensus view assumes an overall growth in consumption. The sustainable-world view assumes radical improvements in efficiency with demand stabilizing after 2000. In the consensus view, coal and oil expand rapidly; in the sustainable-world view, coal contracts and there is a surge in natural gas. Hydropower and commercial renewable fuels would increase by 60 per- cent. Carbon dioxide emissions will increase until 2000 (*right*), but their output will then depend on policies cho- sen in the 1990's. If energy use increases as fast as eco- nomic growth, CO_2 emissions will double by 2010. With greater energy efficiency, emissions in the consensus sce- nario will increase at a rate that is half as fast; in the sustainable view, CO_2 emissions peak after 2000.

barrels in Europe and Japan and more than 40 in the U.S. (see Figure 1.6).

Although reliant on traditional fuels, low-income economies have high energy intensity (energy used per unit of income) because they basically need whatever energy form is available—normally fire- wood, agricultural residues or dung with which to cook and heat their homes—despite the ineffi- ciency with which such fuels are burned. And so a pattern emerges: as countries become increasingly industrialized, the amount of commercial fuel used per unit of income increases, but overall energy intensity declines. Since commercial energy con- sumption may increase as fast as income over long periods, the growth in energy demand in develop- ing countries could be as high as 4 to 5 percent per annum.

Solving energy problems, today as in the past, depends on the technologies that are available and the rate at which they evolve. Since the middle of the 19th century, sources of power have shifted from wind, water and wood to coal and more re- cently to oil and natural gas. The interplay of energy and technology, as exemplified by three phases of the industrial revolution, accounts for the shift.

During the first phase, which emerged in the early 18th century, the dominant technologies were coal mining, the smelting and casting of iron, and steam- driven rail and marine transport. The system's com- ponents intertwined closely: the steam engine, de- veloped originally by Thomas Newcomen for drainage and hoisting in mines, was later adapted by James Watt to provide power for transport and the blast for iron smelters. The smelters, in turn, provided materials for constructing the steam en- gine, locomotives, rails, ships and mining equip- ment. Through the creation of a transportation in- frastructure and the machines to run factories, rapid industrialization was possible.

Toward the end of the 19th century the world was again transformed—this time by electric power, internal-combustion engines, automobiles, airplanes and the chemical and metallurgical indus- tries. Petroleum emerged as a fuel and a feedstock for the petrochemicals industry. Now, toward the end of the 20th century, society has embarked on a third phase of the industrial revolution, character- ized by a shift to computers, advanced materials, optical electronics and biotechnology.

The impact the third phase will have on global

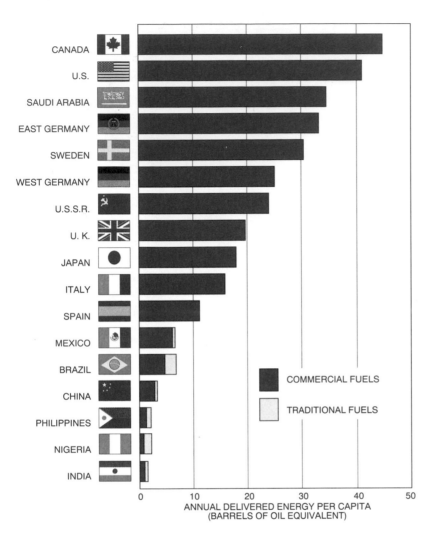

Figure 1.6 GREAT DISPARITIES exist in per capita energy use. The U.S. and Canada have the highest rates: the average citizen in those countries used the equivalent of 40 barrels of oil in 1988. In contrast, the average Nigerian used only two barrels, mostly in the form of traditional fuels. The relatively high rates in the U.S.S.R. and East Germany reflect the inefficiency with which energy there is distributed and used.

patterns of energy consumption is not yet certain, for application of technology depends on what society considers its objectives to be and especially on whether the public will embrace more of a sustainable-world view or not.

Coming to a global consensus will not be easy. Energy policies vary considerably from one country to another. Whereas some governments tax, others subsidize delivered energy. For some, energy is a source of revenue; for others, it is an opportunity to provide relief for the poor. Additional issues that must be faced include the security of supply and the impact of energy price changes on inflation and on a country's balance of trade.

Taxes levied on oil producers, for example, are a mainstay of revenue for the large oil exporters, such as Venezuela and Saudi Arabia. Among countries in the Organization for Economic Cooperation and Development (OECD), it is mainly the consumers of energy who are taxed. Such policies are hardly fixed; indeed, some have undergone radical shifts in a rather short time. In energy-importing countries, concern over security of supply, largely absent until the first oil crisis of 1973, dominated the rest of the 1970's and early 1980's. Later in the 1980's, as surpluses appeared, concern shifted to the other side of the market: oil-exporting countries began seeking secure outlets for their oil.

In many countries the energy sector is the province of state-owned companies; in others, such as the U.K., economic policy has emphasized the virtues of efficiency and competition, thus encouraging privatization. Most governments control the impact on the environment of all aspects of the energy chain—from production to waste disposal to the dismantling and decommissioning of a plant. The common thread in all cases is that a government strives as a matter of policy to provide its people with an adequate, safe, economic and equitable supply.

Implementation of such policies will be profoundly influenced by the evolution of technology, whether applied to the design, operation or control of energy systems. Estimating the rate at which new technologies penetrate the market is problematic, however, because in most cases, the interdisciplinary nature of technological interactions cannot easily be discerned in advance. For example unforeseen synergies between new materials, engineering techniques microelectronic devices and combustion technologies have already greatly increased the potential efficiency with which automobiles consume fuel (see Chapter 5, "Energy for Motor Vehicles," by Deborah L. Bleviss and Peter Walzer), but further increases are possible.

The great potential for saving energy through efficient technology is explored in many chapters in this book. Arnold P. Fickett, Clark W. Gellings and Amory B. Lovins suggested that efficiency measures have the potential to reduce electricity consumption in the U.S. by 30 to 75 percent in Chapter 2, "Efficient Use of Electricity." Rick Bevington and Arthur H. Rosenfeld review various strategies for reducing fuel use in buildings in Chapter 3, "Energy for Buildings and Homes." Similar savings can be achieved in the industrial sector, which currently accounts for 40 percent of the energy used in the developed world (see Chapter 4, "Energy for Industry," by Marc H. Ross and Daniel Steinmeyer).

New highly efficient technologies include compact fluorescent lamps and other lighting devices, which can reduce the amount of electricity required for lights by 90 percent. Also available are appliances that consume only from 10 to 20 percent of the electricity that conventional ones do. A new generation of automated controls makes it possible to optimize lighting, heating, ventilating and air-conditioning systems. In industry, both adjustable-speed drives and high-efficiency motors promise significant savings, as do advances in integrated-process design, control technology and recycling. In the transportation sector, vehicles capable of 60 miles per gallon of fuel or more and vehicles that run on compressed natural gas, hydrogen and electricity are gaining attention.

Developments in exploration and production of oil and gas will also be important. The use of three-dimensional seismic techniques and horizontal drilling, for example, will increase the accessibility of these resources yet not raise their cost significantly. Rapid technological advances in the alternative-energy sector, described by Carl J. Weinberg and Robert H. Williams in Chapter 10, "Energy from the Sun," open up new possibilities. Laboratory efficiencies of solar photovoltaic cells have roughly doubled since the 1970's and are expected to improve further, variable-speed wind turbines are cost-competitive in some markets and new processes for making liquid fuel from biomass suggest viable alternatives to petroleum, although on a smaller scale. Wolf Häfele foresees that the threat of global warming will create a significant demand for nuclear power, one that can be met by inherently safe, diversion-proof reactors managed under the aegis of an international authority in Chapter 9, "Energy from Nuclear Power."

Technology is also mediating a shift away from large, centralized power plants to smaller, decentralized ones. Improvements in electronic communications, control and computing technology have made it easier to monitor and regulate complex grids remotely. With the arrival of new gas turbines, small engines, solar cells and other technologies, the economies of scale, so long a feature of electricity generation, are diminishing. Not only can decentralization be more efficient, but as Amulya K. N. Reddy and José Goldemberg make clear in Chapter 6, "Energy for the Developing World," it may offer some of the poorer countries a basis for economic growth.

In the long run, such technological advances appear likely to lower the overall costs associated with limiting the carbon dioxide emissions from fossil fuels. In general, response times—from concept through prototype to commercial product—are declining, a pattern that is likely to continue. Japanese manufacturers, for example, anticipate that by 1993 cars will be designed and developed in half the time and at a quarter of the cost of today's models. Their approach highlights the role of global competition as a motor of technological change. Since similar advances are expected in other areas of manufacturing, we could be surprised at the speed of the response to new incentives.

Technology, however, will propel society toward sustainability most quickly if policymakers can agree on appropriate global guidelines. If there is to be any significant change in the structure of energy supply and use in the next 20 years, new policies will have to be in place by the mid-1990's, which means that OECD members would have to agree by then on a protocol to offset the possibility of global climate change.

Broadly, the main options are to eliminate chlorofluorocarbons (which deplete stratospheric ozone and contribute to global warming), initiate affores-

tation programs to enlarge the carbon sink and reduce CO_2 emissions from fossil fuels. These goals can be accomplished by improving the efficiency with which fossil fuels are transformed and consumed (see Figure 1.7) and by shifting to alternative fuels, especially from carbon-rich to hydrogen-rich fuels. Technologies to remove carbon dioxide from coal are being developed, although methods for sequestering the CO_2 have yet to be resolved (see Chapter 8, "Energy from Fossil Fuels," by William Fulkerson, Roddie R. Judkins and Manoj K. Sanghvi).

Although each country will undoubtedly differ, certain policies can be expected to gain wide support. Among them is the "polluter pays" principle,

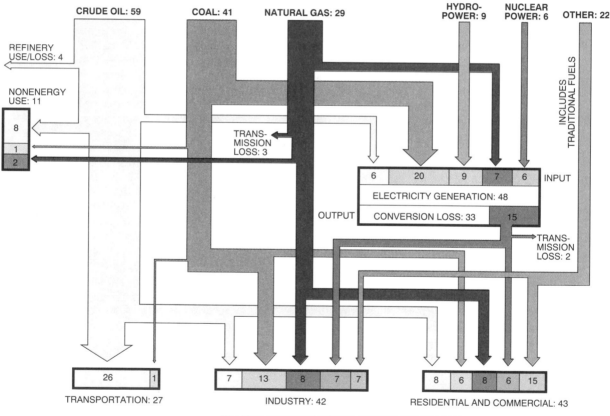

MILLIONS OF BARRELS OF OIL EQUIVALENT PER DAY

Figure 1.7 WORLD ENERGY FLOWS for 1985 show that fossil fuels are highly versatile. Crude oil (*yellow*) must be processed in refineries, where it is converted to gasoline, diesel and aviation fuel for transportation. Both oil and natural gas (*red*) are widely used by industry and by residential and commercial consumers, but only a small percentage of each generates electricity. Most of the world's coal (*light blue*) powers industry or generates electricity. Hydropower, nuclear power and other energy sources (biomass, solar energy and wind power) make up 22 percent of today's primary energy supply.

which calls for the user to pay the full cost of resource use and for the market to act as arbiter of supply and demand. It must also be remembered that CO_2 emissions per capita are 10 times higher in the OECD countries than in developing countries; an equitable agreement on appropriate reductions is therefore necessary.

Involving the developing countries, as well as Eastern Europe, the Soviet Union and China, in the global energy response will be critical. For many of these countries, economic reform is a necessary starting point. As William U. Chandler, Alexei A. Makarov and Zhou Dadi point out in Chapter 7 "Energy for the Soviet Union, Eastern Europe and China," the ongoing globalization of business offers hope for the rapid diffusion of technology from one part of the world to another (see Figure 1.8).

The scope of an effective program has yet to be determined. A recent study, prepared for the 1989 Ministerial Conference on Atmospheric Pollution and Climate Change in Noordwijk, the Netherlands, estimates capital costs of about .8 percent of gross domestic product (GDP) for a prevention program targeting a full phaseout of chlorofluorocarbons, expanded forest management and energy conservation. If fully implemented, such a program could reduce expected greenhouse-gas emissions in OECD countries by 30 percent by 2005. But a full commitment to sustainability could demand more. Appropriate environmental expenditures may need to be increased on the order of 1 or 2 percent of GDP.

Shifts of that magnitude can be carried out over one or two decades with little economic disruption. Similar—even larger—shifts have already occurred. Between 1965 and 1985, for example, the amount spent by U.S. consumers on food declined by 6 percent, whereas health care expenditures increased from 6 to 11 percent during the same period. At the same time, energy expenditures in the

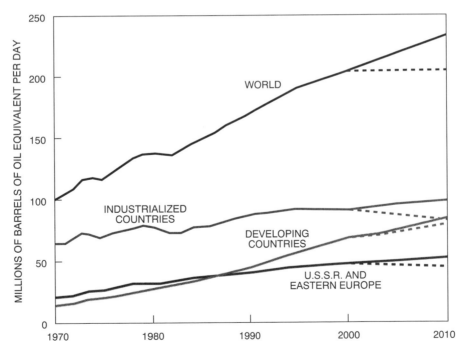

Figure 1.8 PRIMARY ENERGY DEMAND is expected to vary from one part of the world to another. Most of the increase will probably occur in the developing world, where population growth rates are high and industrialization and urbanization are under way. In contrast, demand is expected to remain stable or drop in the industrialized countries, where population growth rates are low. It could stabilize or decline in Eastern Europe and the U.S.S.R. depending on the success of economic reforms. Much hinges on whether consensus or sustainable policies are enacted.

OECD countries dropped from 12 percent of GDP in 1980 to only 8 percent in 1988. Today funding for the environment is poorly documented, but it probably averages 2 or 3 percent of GDP for most OECD countries. Increasing the amount to 4 or 5 percent is certainly feasible (particularly if the defense budgets of many countries decline as expected).

In a world that embraces sustainability and establishes appropriate incentives to trigger change, one could expect the following outcome in the energy sector. Assuming that population and economic growth are the same as in the consensus view and that international protocols are in place by the mid-1990's, the world's primary energy supply could be stabilized at the equivalent of around 205 million barrels of oil a day by 2000. This by itself would be a heroic feat. Yet even with such a program, CO_2 emissions from fossil fuels would be 25 percent higher in 2000 than they are today. Despite a significant shift from coal to natural gas (combined with an increase in renewable fuel supplies and greater efficiency), CO_2 emissions would still be higher in 2010 than they are today.

In a sustainable world, the balance of new initiatives would shift from producer to consumer, from energy supply to energy services and from quantity to quality of energy. An important policy step could be to adopt a number of options that would integrate needed energy services into regional and city planning. As Bleviss and Walzer point out, new zoning laws can deter commuting by car, and new traffic guidance systems might reduce urban air pollution. Bevington and Rosenfeld highlight the value of planting shade trees and painting buildings in light, reflective colors to reduce energy use in urban areas. Changes to the existing transportation infrastructure, such as the proposed introduction of maglev trains in Europe, open up options for rethinking our road, rail and air networks.

The corporate response to a sustainable world could include a new breed of energy company, driven by the desire to provide a broad range of leading energy technologies to its customers. Such companies could have wide-ranging activities. In some urban areas, for example, there may be a market for privately operated mass-transit systems. In others, banning the internal-combustion engine might create facilities for parking and recharging the batteries of electric cars. As utilities grow more service oriented, it is quite possible they will expand to perform such services. The need to move quickly should also encourage research and development alliances between fuel companies and manufacturers of combustion equipment. Such collaborations can produce fuels, engines and processes that are as yet unanticipated. A modest start has already been made in the U.S., where some petroleum companies are producing reformulated gasoline and alternative fuels.

For many, the transition to a sustainable world is rife with uncertainty and dilemmas. This is perhaps natural since we seem to be living between two stories, as expressed by the consensus and the sustainable-world views. For others, the situation is more clear-cut. As stated so clearly in Managing Planet Earth: Readings from *Scientific American Magazine* (1989), for them, the story of the sustainable world is the story of our time, a time when human beings need to reaffirm their role as stewards (see Chapter 11, "Energy in Transition," by John P. Holdren).

As we learn more about the relation human beings have to their planet, we may find that rather than viewing energy as a commodity to be exploited *from* planet Earth, we will increasingly need to think and act in terms of energy *for* planet Earth. Our dependence on energy will persist, but it must do so in the context of an ecologically sound planet. This means human beings may well have to apply all their inventiveness to develop new energy technologies so as to guarantee the long-term quality of their habitat.

Efficient Use of Electricity

Advanced technologies offer an opportunity to meet the world's future energy needs while minimizing the environmental impact. Both suppliers and consumers of electricity can benefit from the savings.

. . .

Arnold P. Fickett, Clark W. Gellings and Amory B. Lovins

Electricity is fundamental to the quality of modern life. It is a uniquely valuable, versatile and controllable form of energy, which can perform many tasks efficiently. In little over 100 years electricity has transformed the ways Americans and most peoples of the world live. Lighting, refrigeration, electric motors, medical technologies, computers and mass communications are but a few of the improvements it provides to an expanding share of the world's growing population.

Many analysts believe that regional electricity shortages could occur in the U.S. within the next 10 years, perhaps as early as 1993. Given the importance of electricity to all sectors of the economy, such shortages would have severe consequences. Yet financing large-scale power plant construction could push America's $170-billion-a-year electricity costs higher: a large (one billion watts) power plant costs more than $1 billion and may entail lengthy regulatory and environmental approvals. Thus there is growing pressure for utilities to provide needed generating capacity or to reduce electricity demand, or both.

A kilowatt-hour of electricity can light a 100-watt lamp for 10 hours or lift a ton 1,000 feet into the air or smelt enough aluminum for a six-pack of soda cans or heat enough water for a few minutes'

shower. To save money and ease environmental pressures, can more mechanical work or light, more aluminum or a longer shower be wrung from that same kilowatt-hour?

The answer is clearly yes. Yet estimates as to how much more range from 30 to 75 percent. Also at issue is how fast efficiency can be improved and at what cost (see Figure 2.1).

Since the oil embargo of 1973, energy intensity—the amount of energy required to produce a dollar of U.S. gross national product—has fallen by 28 percent. Plugged steam leaks, caulk guns, duct tape, insulation and cars whose efficiency has increased by seven miles per gallon have helped to extract more work from each unit of fuel. Applications of electricity, too, have made important contributions to productivity and to a more information-based economy. Electricity accounts for a growing fraction of energy demand, and its relation to the gross national product has held relatively steady in recent years. It is not clear, however, that electricity and economic growth must continue to march in lockstep. Technologies and implementation techniques now exist for using electricity more efficiently while actually improving services. Harnessing this potential could get society off the present treadmill of ever higher financial and environmental risks and

could make affordable the electric services that are vital to global development (see Figure 2.2).

Historical patterns are already starting to change. California reduced its electric intensity by 18 percent from 1977 to 1986 and expects the trend to continue. Nevertheless, in such major industries as cars, steel and paper, Japan's electric use per ton is falling while the U.S.'s is rising—chiefly because American companies are still adopting new fuel-saving "electrotechnologies" already common in Japan. But companies there are improving their efficiency at a faster rate. The resultant widening efficiency gap contributes to Japanese competitiveness.

Other industrialized nations are also setting higher standards for efficiency. Sweden has outlined ways to double its electricity efficiency. Denmark has vowed to cut its carbon dioxide output to half the 1988 level by 2030 and West Germany to 75 percent of the 1987 level by 2005; both nations emphasize efficiency.

These encouraging developments reflect rapid progress on four separate but related fronts: advanced technologies for using electricity more productively; new ways to finance and deliver those technologies to customers; expanded and reformulated roles for electric utilities; and innovative regulation that rewards efficiency.

The technological revolution is most dramatic. The 1980's created a flood of more powerful yet cost-effective electricity-saving devices. If anything, progress seems to be accelerating as developments in materials, electronics, computer design and manufacturing converge. Rocky Mountain Institute estimates that in the past five years the potential to save electricity has about doubled, whereas the average cost of saving a kilowatt-hour has fallen by about two thirds. The institute has also found that most of the best efficient technologies are less than a year old.

Of course, while some innovations are saving electricity, others will use electricity in new ways in those areas where electricity has an advantage over other forms of energy. For example, electricity can be environmentally beneficial and cost-effective in ultraviolet curing of finishes, microwave heating and drying, induction heating and several other industrial uses. Such electrotechnologies save money and fuel and reduce pollution overall. The Electric Power Research Institute (EPRI) estimates that by 2000 these new technologies will save as much as half a quadrillion British thermal units (Btu's) of fuel per year yet will increase electricity use in the U.S. only slightly.

How much electricity could be saved if we did everything, did it right and fully applied the best technologies for efficiency? Agreement is growing that an astonishing amount of electricity—far more than the 5 to 15 percent cited a few years ago—could be saved in the U.S. (see Figure 2.3). According to a 1990 report by EPRI, it is technically feasible to save from 24 to 44 percent of U.S. electricity by 2000—some of it rather expensively—in addition to the 9 percent already included in utility forecasts. Thus, theoretically, aggressive efficiency efforts might capture as much as three to five times the savings that EPRI forecasts to happen spontaneously, about four to seven times as much as current utility programs plan to capture (80 billion watts before 2000). Rocky Mountain Institute estimates a long-term potential to save about 75 percent of electricity at an average cost of .6 cent per kilowatt-hour—several times lower than just the cost of fuel for a coal or nuclear plant. Even more could be saved at higher costs. The differences between these estimates are less important than their agreement that substantial amounts of electricity can be saved in a cost-effective manner.

How do potential electricity savings in the U.S. compare with analyses for other countries? Potential savings vary, mainly because of differences in climate, in use of appliances and in price and economic structure. Western Europeans and the Japanese have already captured more of the potential electricity savings, and, as these nations continue to progress, they will pay more for less electricity savings than Americans, but the differences are probably not substantial. Studies have found potential savings of 50 percent in Sweden at an average cost of 1.3 cents per kilowatt-hour, 75 percent in Danish buildings at 1.3 cents per kilowatt-hour and 80 percent in West German households at a cost repaid in 2.6 years (see Figure 2.4).

Strong anecdotal evidence suggests that in most developing and socialist countries, many electric devices are several times less efficient than in the U.S. Improved devices are often costly there today be-

Figure 2.1 ELECTRICITY carried by power lines in Torrance, Calif., runs appliances, heats homes and lights buildings. Technology can improve these services and at the same time save electricity and money.

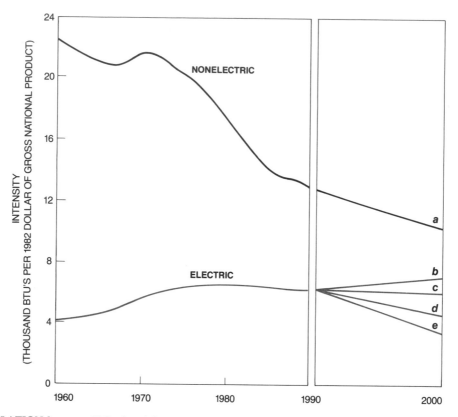

Figure 2.2 RELATION between U.S. electricity consumption and the economy has remained relatively steady over the past three decades, whereas the intensity of other forms of energy has declined, partly because of the increased use of electricity. Efficient use of electricity, however, can reduce intensity. The graph shows a projection of nonelectric intensity (*a*), a projection of electric intensity at current efficiency levels (*b*), a projection that also includes utility efficiency programs (*c*), a conservative estimate of efficiency potential (*d*) and an optimistic estimate of the potential (*e*).

cause they require electronics or special materials that are not readily available. But as global markets for these devices expand, lowering their international prices, it is reasonable to expect that the potential electricity savings will be even greater in the countries that are the least efficient today. The U.S. potential may therefore prove to be not a bad surrogate for the global average.

To understand the pitfalls involved and the effort required to move toward a more efficient economy, consumers and suppliers of electricity must understand how major savings can be achieved. Electricity, like other forms of energy, can be saved by demanding fewer or inferior services—warmer beer, colder showers, dimmer lights. No such options are considered here. If technology is applied

intelligently, electricity can be saved without sacrificing the quality of services. In fact, many new devices actually function better than the equipment they replace: they provide more pleasing light, more reliable production and higher standards of comfort and control.

The biggest savings in electricity can be attained in a few areas: lights, motor systems and the refrigeration of food and rooms. In the U.S. lighting consumes about a quarter of electricity—about 20 percent directly, plus another 5 percent in cooling equipment to compensate for the unwanted heat that lights emit. In a typical existing commercial building, lighting uses about two-fifths of all electricity directly or more than half including the cool-

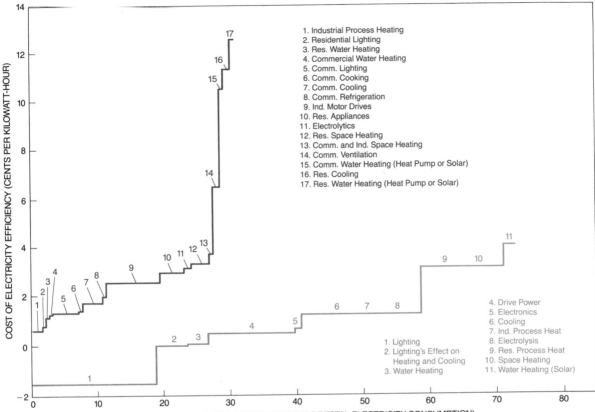

1. Industrial Process Heating
2. Residential Lighting
3. Res. Water Heating
4. Commercial Water Heating
5. Comm. Lighting
6. Comm. Cooking
7. Comm. Cooling
8. Comm. Refrigeration
9. Ind. Motor Drives
10. Res. Appliances
11. Electrolytics
12. Res. Space Heating
13. Comm. and Ind. Space Heating
14. Comm. Ventilation
15. Comm. Water Heating (Heat Pump or Solar)
16. Res. Cooling
17. Res. Water Heating (Heat Pump or Solar)

4. Drive Power
5. Electronics
6. Cooling
7. Ind. Process Heat
8. Electrolysis
9. Res. Process Heat
10. Space Heating
11. Water Heating (Solar)

1. Lighting
2. Lighting's Effect on Heating and Cooling
3. Water Heating

(axes: COST OF ELECTRICITY EFFICIENCY (CENTS PER KILOWATT-HOUR); POTENTIAL ELECTRICITY SAVINGS (PERCENT OF TOTAL ELECTRICITY CONSUMPTION))

Figure 2.3 EFFICIENT TECHNOLOGIES offer the potential to reduce long-term U.S. electricity consumption as estimated by the Electric Power Research Institute (*red line*) and by Rocky Mountain Institute (*blue line*). Estimates are given in 1990 dollars.

ing load. Converting to today's best hardware could save some 80 to 90 percent of the electricity used for lighting, according to Lawrence Berkeley Laboratory. EPRI suggests that as much as 55 percent could be saved through cost-effective means.

Compact fluorescent lamps, for instance, consume 75 to 85 percent less electricity than do incandescent ones. They typically last four to five times longer than incandescent floodlamps and nine to 13 times longer than ordinary incandescent bulbs. If one balances the higher initial cost of the lamps against the reduction in replacement lamps and installation labor (longer-life bulbs do not need to be changed so often), one can recover the cost of the fluorescent lamps and still save many dollars over the life of each lamp. One can thus make money

without even counting the savings in electricity. This is not a free lunch; it is a lunch you are paid to eat.

Efficient lighting hardware is now available for almost any application. Most devices provide the same amount of light as older systems do, with less glare, less noise, more pleasant color and no flicker. These aesthetic improvements can unlock even bigger savings: improving productivity by 1 or 2 percent is usually worth more to an office's bottom line than eliminating electric bills.

Together the lighting innovations that are commercially available can potentially save one-seventh to one-fifth of all the electricity now used in the U.S. These innovations would cost about one cent per kilowatt-hour to install. The reduced mainte-

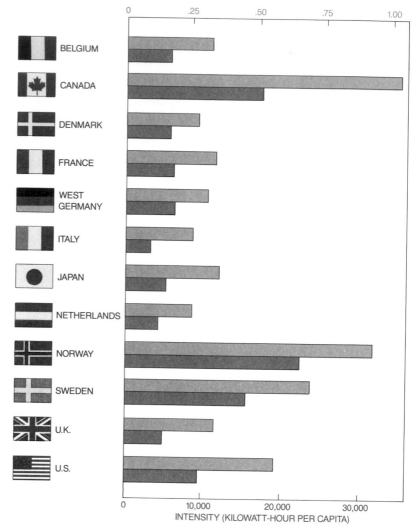

Figure 2.4 VARIATIONS in the habits of many developed nations suggest that each one has considerable flexibility in choosing strategies for consuming and supplying electricity. Blue bars indicate electricity consumption per 1986 dollar of gross domestic product; red bars show electricity consumption per person in 1986.

nance costs, Rocky Mountain Institute calculates, would save the user an additional 2.4 cents per kilowatt-hour saved. The savings in electricity would reduce the need for about 70 to 120 billion watts of power plants that would cost from $85 to $200 billion to build and from $18 to $30 billion per year to operate. Thus, lighting innovations may be the biggest gold mine in the entire economy (see Figure 2.5).

After lights, electric motors offer the best opportunity to effect major savings. Motors consume 65 to 70 percent of industrial electricity and more than half the electricity generated in the U.S. The annual

electricity bill for motors exceeds $90 billion, or about 2 percent of the gross national product.

A typical large industrial motor consumes electricity that costs some 10 to 20 times its total capital cost per year: its capital cost is only about 1 to 3 percent of its total life-cycle cost. Over a motor's life, a one-percentage-point gain in efficiency typically adds upward of $10 per horsepower to the bottom line. Multiplying the many percentage points of available savings by the hundreds or thousands of horsepower for each big motor reveals very large potential savings. And one industrial processing plant can have hundreds of such motors (see

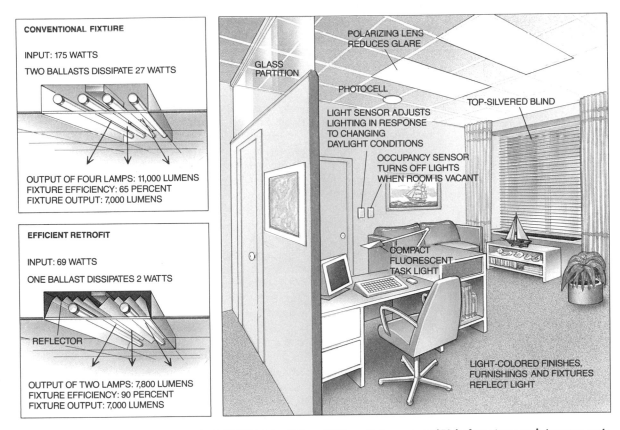

CONVENTIONAL FIXTURE

INPUT: 175 WATTS

TWO BALLASTS DISSIPATE 27 WATTS

OUTPUT OF FOUR LAMPS: 11,000 LUMENS
FIXTURE EFFICIENCY: 65 PERCENT
FIXTURE OUTPUT: 7,000 LUMENS

EFFICIENT RETROFIT

INPUT: 69 WATTS

ONE BALLAST DISSIPATES 2 WATTS

REFLECTOR

OUTPUT OF TWO LAMPS: 7,800 LUMENS
FIXTURE EFFICIENCY: 90 PERCENT
FIXTURE OUTPUT: 7,000 LUMENS

GLASS PARTITION

POLARIZING LENS REDUCES GLARE

PHOTOCELL

LIGHT SENSOR ADJUSTS LIGHTING IN RESPONSE TO CHANGING DAYLIGHT CONDITIONS

TOP-SILVERED BLIND

OCCUPANCY SENSOR TURNS OFF LIGHTS WHEN ROOM IS VACANT

COMPACT FLUORESCENT TASK LIGHT

LIGHT-COLORED FINISHES, FURNISHINGS AND FIXTURES REFLECT LIGHT

Figure 2.5 1.5 BILLION LIGHTING FIXTURES in U.S. buildings could use about 60 percent less electricity—typically from 70 to 90 percent less when using dimming technologies available in 1988. Lamp output is improved by phosphors, cooler fixtures and operation at high frequency (about 30 kilohertz). The retrofit costs less than $130 per fixture, saves $50 in long-term maintenance costs and pays for itself in one or two years. It saves electricity at a net cost of about .6 cent per kilowatt-hour. The further options shown can save even more, approaching 100 percent in new specially daylit buildings.

Chapter 4, "Energy for Industry," by Marc H. Ross and Daniel Steinmeyer).

Many machines, especially pumps and fans, need to vary their output to accommodate changing process needs. This is often done by running the pump or fan at full speed while "throttling" its output with a partly closed valve or damper—like driving with one foot on the gas and the other on the brake. Electronic adjustable-speed drives can reduce this waste. When you need only half the flow from a pump, you can in principle save seven-eighths of its power and in practice nearly that much, because electronic drives are very efficient. Savings can range from 10 to 40 percent, with typical applications reducing total U.S. motor energy by about 20 percent. Paybacks range from six months to three years, averaging one year.

A new breed of high-efficiency motors represents another important advance. These motors are better designed and better made from higher-quality materials than conventional motors are, thereby squeezing down their magnetic, resistive and mechanical losses to less than half the levels of a decade ago. Although such motors are widely found in North America, Western Europe and Japan, they are not available in some industrialized countries, such as Australia, or in much of the socialist or developing world, where motor efficiencies are often very low.

Most engineers think only in terms of adjustable-speed drives and high-efficiency motors. Although both are important, they account for only half of the total potential electricity savings in U.S. motor systems. The other half comes from 33 other improvements in the choice, maintenance and sizing of motors, three further kinds of controls and the efficiency with which electricity is supplied to the motor and torque is transmitted to the machine. Improved motor systems can run on about half the electricity, which amounts, in principle, to electricity savings equivalent in the U.S. to about 80 to 190 billion watts of power plants. Furthermore, the cost of the new motors can usually be recovered in about 16 months, because after a company pays for only seven of the 35 improvements (including adjustable-speed drives and high-efficiency motors), the other improvements are free by-products.

Progress in motor and lighting technologies is matched by advances in superefficient appliances. Refrigerators and freezers can now consume 80 to 90 percent less electricity than conventional models; commercial refrigeration systems can save 50 percent, televisions 75 percent, photocopiers 90 percent and computers 95 percent. Rocky Mountain Institute, for example, installed efficient lights and appliances in its 4,000-square-foot headquarters, decreasing consumption of electricity tenfold, to only $5 a month. The institute also lowered its water consumption by 50 percent and eliminated 99 percent of the energy it needed to heat space and water. (The building is so well insulated that even though it is located in the subarctic climate of the Colorado mountains it needs no furnace). Moreover, all of this seven-year-old technology paid for itself within a year.

Exploiting the full menu of efficiency opportunities can double the quantity and more than halve the cost of savings, because saving electricity is like eating a lobster: if you extract only the large chunks of meat from the tail and claws and throw away the rest, you will miss a comparable amount of tasty morsels tucked in crevices. To capture major electricity savings cheaply, one must not only install new technologies but also rethink the engineering of whole systems, paying meticulous attention to detail.

Such rethinking will require a new infrastructure to deliver integrated packages of modern technologies. Only a handful of firms provide comprehensive, up-to-date lighting retrofits; few if any provide similar services for motors (see Figure 2.6). Yet retrofits that save electricity represent a global business opportunity ultimately worth perhaps hundreds of billions of dollars a year—an ample prize to elicit entrepreneurship (see Chapter 3 "Energy for Buildings and Homes," by Rick Bevington and Arthur H. Rosenfeld).

The potential to save electricity will not be realized until—like power plants—electricity-saving programs are planned, designed, financed, built, commissioned and maintained. Just as one might extract a mineral resource from the ground, one must determine how much electricity can be profitably saved employing existing technology and how to convert that reserve to actual production.

Efficient technologies are often underused because of the lack of customer demand (market pull) or the lack of a sufficient distribution channel (market push), or both. If electricity consumers want efficient appliances and ask retailers to provide them, retailers will then ask wholesalers to supply them, and wholesalers in turn will seek manufacturers to produce those products. If consumers fail to act, then the whole string of potential benefits unravels.

To create market pull, energy planners must understand how consumers make energy choices. Most planners are puzzled to find that customers sometimes shun efficiency even when it is accompanied by attractive economic incentives. In the past, manufacturers and retailers have not considered efficiency to be an important feature in new products, because they have found that consumers rarely decide to make a purchase based on efficiency. The factors that most consistently affect their choices are appearance; safety; comfort, convenience and control; economy and reliability; high-technology features; the need to have the latest equipment; the desire to avoid hassles; and resistance to having utilities control energy use. Because human nature is diverse, the weighing of these factors varies enormously, and retailers must adjust their marketing strategies accordingly. Businesses have analogous concerns, including product quality, production reliability, fuel flexibility, environmental cleanliness, a clean workplace and low risk.

If efficient technologies are to be widely deployed, a third party, such as the electric utility or government, may need to assume responsibility for both market push and market pull. As we shall see, utilities have a special interest in influencing customers' demand—treating it not as fate but as choice—in

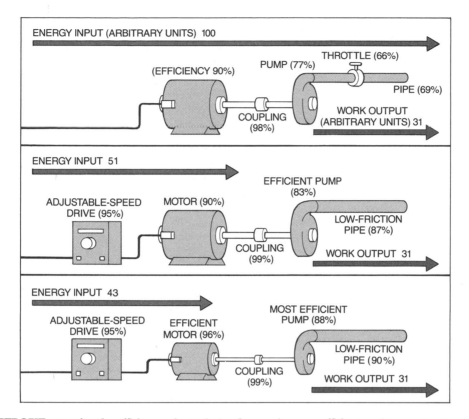

Figure 2.6 RETROFIT can raise the efficiency of a typical motor-pump system (*top*) from 31 to 72 percent and can pay for itself in two to three years (or less counting saved maintenance costs). An electronic drive (*middle*) affects the efficiency of the other components. Here the drive's net effect is a 21 percent savings, not counting lower pipe **losses. A more efficient and properly sized motor and pump, as well as better pipes, saves even more (*bottom*), partly by fixing the damage caused by improper repairs to the old motor. Further refinements may cut energy use by 40 percent.**

order to provide better service at lower cost while increasing their own profits and reducing their business risks. Utilities can choose from a wide range of market push and pull methods designed to influence consumer adoption and reduce barriers. These include rebates or other financing options, direct contact with their customers, special tariffs, advertising, education, and cooperative ventures with architects, engineers and suppliers of efficient technology. Collectively, such efforts are part of demand-side management, which seeks to change the demand for electricity while still meeting customers' needs.

More than 60 utilities serving almost half of all Americans now offer rebate programs to promote the buying or selling of efficient devices. The over-

whelming majority (92 percent) pay rebates to purchasers to create market pull; about 24 percent pay appliance dealers to create market push.

Utility rebate programs can rapidly stimulate market development. Efficient lighting equipment was unavailable in Las Vegas, for instance, until Nevada Power Company started offering rebates, whereupon within six months, 20 wholesale and retail outlets were competing in the price and breadth of efficient lighting systems.

Many utilities have begun to pay consumers for each kilowatt-hour saved, no matter how it is done. They have also tried to reward "trade allies" who remove old, inefficient equipment or who sell, specify or install electricity-saving devices. Utilities sometimes offer rebates to consumers who beat a government performance standard, thus eliciting better technologies so the standard can be raised until cost-effectiveness limits are reached.

Other financial incentives complement rebates: low- or no-interest loans, gifts and leases. Southern California Edison Company, for example, has given away more than 800,000 compact fluorescent lamps. The Taunton Municipal Lighting Plant in Massachusetts leases such lamps for 20 cents each per month and replaces them for free. Thus, customers can pay for efficiency over time, just as they would otherwise pay for power plants. The makers of compact fluorescent lamps have relied on both their own and utilities' marketing strategies to achieve annual U.S. sales of about 20 million units. Those sales are doubling or tripling each year, and such lamps already dominate the West German market.

These well-established methods are so effective that when Southern California Edison Company had a peak load of 15 billion watts, in 1983–1984, it was able to reduce its forecast of peak demand by more than 500 million watts in a single year. At the same time, California's appliance and building standards increased electricity savings even more. Annual savings represented 8.6 percent of the utility's peak demand at the time and cost the utility only about 1 percent as much as building and running a new power station. If all Americans saved electricity as fast as those 10 million did, the U.S. economy could grow by several percent every year while total electricity use decreased.

Such success stories are now spreading in the U.S. and abroad. In some instances, skillful and imaginative marketing has captured 70 to 90 percent of specific efficiency markets, such as housing insulation, in just a year or two. Some utilities, such as the Bonneville Power Administration, are saving businesses money through commercial efficiency programs whose cost is about .5 cent per kilowatt-hour.

Utilities such as North Carolina's Duke Power Company offer lower rates to efficient customers. Others require minimum efficiency levels as a condition of service; Atlantic Electric in New Jersey, for example, has such an air conditioner standard. Several states are now trying or considering a sliding-scale hookup fee: when a utility connects a new building to the power grid, it charges a fee that is tied to the building's efficiency. Consideration is also being given to using such fees to pay rebates ("feebates") for the most efficient buildings.

Still further savings may be achieved by methods that seek not merely to market "negawatts" (saved electricity) but to make markets in negawatts: saved electricity can be treated as a commodity just like copper or sow-bellies. This strategy can maximize competition among means of savings and among providers of savings and so drive down the cost. For example, some utilities run competitive bidding processes in which all ways to make or save electricity compete.

Saved electricity can be converted to money and traded between utilities or between customers. Some utilities may even want to become "negawatt brokers" and make spot, futures and options markets in saved electricity. Others are considering buying contracts from their customers to stabilize or reduce demand. The contracts could be resold in secondary markets, just as some brokers already buy and sell air pollution rights.

Some aggressive utilities competing in the emerging negawatt market even sell efficiency in the territories of other utilities. Puget Sound Power and Light Company sells electricity in one state, but its subsidiary sells efficiency in nine states.

E ven though some utilities and consumers have taken the lead in electricity efficiency, most of the potential savings remain untapped. Customers use very different financial criteria to assess ways to save electricity than utilities use to assess new power plants. On the one hand, if customers invest money to save electricity in their home or business, they will probably want to recoup their investment within about two years—perhaps as long as five years for a few far-sighted industries and less than one year for low-income renters. On the other hand, if utilities build plants to increase capacity, their technical and financial strength lets them recover costs over a 20-year period.

The gap between the payback horizon of consumers and utilities tends to make society buy too little efficiency and too much supply. The result in the U.S. alone is the $60 billion per year now spent in expanding electricity supplies that could be partly displaced by investments in efficiency. The payback gap also dilutes price signals. If customers can avoid a tariff of six cents per kilowatt-hour by saving electricity, then without other incentives they will buy efficiency costing up to .6 cent per kilowatt-hour—about a tenth of the tariff, because the tariff is calculated at the utility's payback horizon of 20 years, but the customer invests on the basis of a two-year horizon. Just getting the prices right will therefore not necessarily induce people to buy as much efficiency as would benefit society at large.

However, correct pricing is important: only prices that tell the truth can inform customers about how much is enough. Prices should be adjusted to the time and season of use—perhaps ultimately with sophisticated new kinds of electronic meters—and reflect real-time spot prices in order to provide the most accurate signals.

Utilities around the world are reexamining their purpose. Is their mission the production and sale of electricity, or is it the profitable production of customer satisfaction? Utilities that take the latter view believe that if electricity costs more than efficiency, then customers will eventually realize they can save kilowatt-hours and money and still get hot showers and cold beer with high-performance shower heads and superefficient refrigerators. The only relevant question, then, is who will sell efficiency? If efficiency is cheaper than electricity; customers will buy less electricity and more efficiency. It is generally a sound business strategy to satisfy customer needs before someone else does.

Utilities are the logical organizations to expedite the use of energy-efficient products: they have technical skill, permanence, credibility, close ties to customers, a relatively low cost of capital and a fairly steady cash flow. At present, however, they have little motive to expedite energy efficiency. The conflict is obvious: Why spend money to reduce sales?

In principle, utilities can profit in several ways from making their customers more efficient. They can avoid operating costs in the short run, construction costs of new power plants in the medium run and replacement costs of old power plants in the long run. They can also earn a spread on financing efficiency, just as a bank would. Legislation such as the amended U.S. Clean Air Act may allow utilities to use efficiency to generate pollution rights, which they can resell. And finally, under new regulations now being adopted in some states in the U.S., utilities may be able to receive exemplary financial rewards for money-saving investments.

A major breakthrough occurred in 1989 when new regulations were accepted in principle nationwide for consideration by state regulators. The proposed rules would uncouple utilities' profits from their sales, removing a utility's disincentive to invest in efficiency. In effect, the utilities will be compensated for the revenue they would otherwise lose by selling less electricity—and will get to keep part of the savings.

Such rules have already proved effective in a few cases. Pacific Gas & Electric Company in California and a group of environmentalists, government administrators and consumers recently agreed that the utility should keep 15 percent of any money saved by certain new efficiency programs. Customers will benefit by getting 85 percent rather than all of nothing.

In New York Niagara Mohawk Power Corporation has proposed another way to profit from efficiency services. Under the plan, the utility's 12 efficiency programs, which cost $30 million to implement in 1990, will be allowed to recover costs and clear a $1-million profit if the utility's 12 programs achieve the state's goal of saving 133 million kilowatt-hours, which is worth about $10 million a year in reduced energy cost for participating customers. By 1992 the programs should save 240 million kilowatt-hours per year. Where does the money come from? Prices per kilowatt-hour will rise by as much as 1.4 percent, yet participating customers will still pay lower bills because they will consume less.

Niagara's residential low-cost measures program, for example, provides each participating household with a low-flow shower head, a compact fluorescent light bulb and insulation to wrap their electric water heaters and pipes. The equipment should save 960 kilowatt-hours per participating household per year. For each household, the utility loses about $72 in annual energy sales but saves about $40 on fuel and capacity costs. The difference ($32 a year) is charged to the residential customers each year for eight years and includes a $5 profit for the utility. For the equipment, each participating household pays $6 a year for eight years. Therefore, each household will save $272 over eight years.

As efficient technologies and implementation techniques spread, how will they change the economics of our businesses, the services we receive and the health of our environment? Consider first the effect of efficiency on local business. In Osage, Iowa (population 4,000), a utility manager launched a nine-year program to weatherize homes and control electricity loads at peak periods. These initiatives saved the utility enough money to prepay all its debt, accumulate a cash surplus and cut inflation-corrected rates by a third (thereby attracting two factories to town). Furthermore, each household received more than $1,000 of savings a year, boosting the local economy and making shops noticeably more prosperous than in comparable towns nearby. If other communities in the U.S. followed the lead of Osage, they could create economic vitality that would reverberate from Main Street to Wall Street.

Electric efficiency can also enhance industrial competitiveness. When the rod, wire and cable business fell on hard times around 1980, for example, the biggest independent U.S. firm, Southwire, responded by saving, over eight years, about 60 percent of its gas and 40 percent of its electricity per ton of product. The savings yielded virtually all the company's profits during a tough period. The efforts of two engineers may have saved 4,000 jobs at 10 Southwire plants in six states.

Electric efficiency could also break a major logjam in global development. In developing nations, electricity generation already consumes a fourth of global development capital, and in the next few decades the utilities of those nations are projected to need about eight times more capital than is expected to be available—a prescription for power shortages. But efficiency can be the key to saving the capital desperately needed for other development tasks.

Electric efficiency can also ease environmental pressures. If a consumer replaces a single 75-watt bulb with an 18-watt compact fluorescent lamp that lasts 10,000 hours, the consumer can save the electricity that a typical U.S. power plant would make from 770 pounds of coal. As a result, about 1,600 pounds of carbon dioxide and 18 pounds of sulfur dioxide would not be released into the atmosphere, reducing the contribution of these gases to global warming and acid rain. Alternatively, an oil-fired electric plant would save 62 gallons of oil—enough to fuel an American car for a 1,500-mile journey. Yet far from costing extra, the lamp generates net wealth and saves as much as $100 of the cost of generating electricity. Since saving the fuel is cheaper than burning it, environmental problems can be abated at a profit. (Power plants that run on fossil fuel use three units of fuel to make one unit of electricity, whereas in socialist and developing countries they often use five to six units to do the same.)

No matter how electric efficiency is used to reduce emissions, consumers and suppliers of electricity will achieve the biggest reduction at the lowest cost in the shortest time only if they choose the best buys first. Suppose a government wants to reduce carbon dioxide emissions by reducing the amount of electricity generated by coal-fired power plants. To replace that electricity, the government should invest in low-cost efficiency options such as lighting or motor retrofits before considering alternative high-cost technologies such as solar or nuclear power. Otherwise each dollar spent will replace less coal burning than it could have. As we compete for limited resources, the order of environmental priority should be the order of economic priority.

The best-buys-first sequence can be determined either by "least-cost utility planning" or "integrated resource planning"—a formal procedure now required by utility regulators in most of the U.S.—or by an equivalent market process in which all ways to make or save electricity compete fairly for marginal investment.

Electric efficiency, wisely bought today, can go far to stretch the electricity supply. It can also provide time to perfect and deploy renewable energy resources such as solar power, an area where recent progress has been so encouraging. If efficiency decreases the demand for electricity, then renewable resources can be deployed more easily and provide more electricity to more people. Both in the broad sense and in detailed design, electric efficiency and renewable resources are natural partners.

The electric utility is only one of many organizations that should be encouraging energy efficiency. State and local agencies can be particularly helpful in educating customers. Federal support for such programs, which were largely abandoned over the past decade, should be restored.

America's largest landlord—the U.S. government—can take the lead by starting a massive, modern retrofit program in federally owned buildings. The government could be the key to developing market push in certain technologies. It could provide funds to help underwrite the high initial manufacturing costs that penalize new technologies. In addition, state and federal authorities could encourage manufacturers to make more efficient products and broaden performance labeling (see Figure 2.7).

Governments could also do more to assist in the research and development of efficient technology. Investments in efficiency are far out of line with potential benefits. Not only do consumers and suppliers of electricity need more and better hardware choices, but they also need better ways to help designers choose from the bewildering large array of technologies that are already available.

A formidable challenge to electric utilities and governments, then, as well as to customers, design professionals and many other stakeholders, is to integrate the technical, economic, cultural, market-

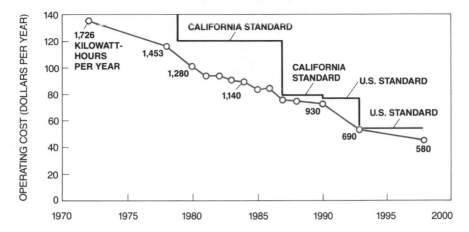

Figure 2.7 GOVERNMENT STANDARDS have pushed manufacturers of refrigerators to produce more efficient models. The average new refrigerator in 1990 costs about $64 less per year to run than one made in 1970. The data were compiled by the Association of Home Appliance Manufacturers and Lawrence Berkeley Laboratory.

ing and policy innovations into coherent efforts to capture the efficiency potential. It is encouraging that many are rising to this challenge. The seriousness of some U.S. utilities' effort, such as that of the New England Electric System, is indicated by their commitment to allocate as much as 4 percent of their gross revenues to improving customers' end-use efficiency. In recent weeks, five U.S. utilities have added nearly $1 billion to their efficiency budgets. Some utilities in Western Europe and Japan, too, have undertaken similarly impressive programs. With such efforts, electric and economic growth need not march in lockstep—if we choose to use electricity in a way that saves money and the environment.

Energy for Buildings and Homes

New technologies—superwindows, compact fluorescent lights and automated-control systems—combined with other strategies, such as shade trees and light-colored buildings, could reduce building energy bills by half.

. . .

Rick Bevington and Arthur H. Rosenfeld

In the moderately populated but energy-intensive U.S., buildings consume 36 percent of the country's energy supply and each year run up an energy bill of nearly $200 billion. Commercial buildings (offices, stores, schools and hospitals) alone have an annual bill of $80 billion (see Figure 3.1). Aside from being expensive, energy takes a tremendous toll on the environment; the energy that powers our appliances and heats, cools and lights our buildings produces 500 million tons of carbon dioxide a year, or two tons of carbon for every person in the U.S.

Yet studies show that the energy efficiency of buildings could double by 2010, cutting carbon emissions in half and saving $100 billion a year, money that could then be poured into economic growth. In fact, significant advances in the building sector have already been made, many of them sparked by the oil crises of the 1970's. During that time, the upward trend in energy use in homes and buildings fell some 30 percent (see Figure 3.2). But in 1986, when the price of oil fell dramatically, the

movement toward greater efficiency stalled. And today although building improvements continue, they are doing so at a slower rate; as a result, energy use in the U.S. building sector is once again on the upswing and is growing at the rate of 3.3 percent a year.

What explains the current trend? To begin with, most building occupants view energy as intangible and automatic, remembering it only when the monthly energy bill arrives. Consumers rarely associate any of the services energy provides—refrigeration, lighting and heat—with the coal mines, oil rigs and power plants that supply this energy. Yet buildings, unlike cars (which are traded in every 10 years or so), tend to last from 50 to 100 years. Over the life of a building, therefore, energy-related improvements make tremendous economic sense.

Rather than being deterred by the costs of energy improvements, consumers should view them as attractive investment opportunities. This can be demonstrated by comparing the cost of two new homes in Chicago. One has conventional walls, with standard 3.5-inch insulation and an annual heating bill of $200. The other has thicker walls and six inches of insulation. Although the extra insulation cost $300 to install, the heating bill for the more efficient house drops to only $80 a year. Thus, the payback period for the insulation is about 2.5 years, which

Figure 3.1 COMMERCIAL BUILDINGS harbor tremendous potential for energy savings. Replacing the lights and the heating, ventilating and air-conditioning (HVAC) systems with new equipment can cut a building's energy costs by 30 percent.

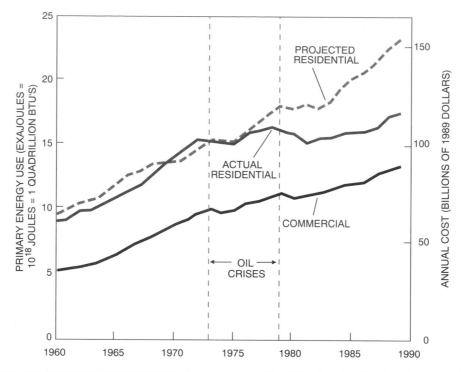

Figure 3.2 RATE OF PRIMARY ENERGY USE (in exajoules) in U.S. buildings has slowed since 1973. Before then energy use in homes and buildings was increasing at annual rates of 4.5 and 5.4 percent, respectively. From 1973 through 1986 energy growth in the domestic sector stopped; in the commercial sector it slowed to 1.6 percent a year. The 1986 drop in oil prices, however, is once again sending the growth curve climbing at a rate of about 3.3 percent a year. Primary energy for electricity refers to the amount used at the power plant, not delivered to consumers.

translates into a 40 percent annual return on an investment of $300. Unfortunately, homeowners rarely consider the advantages of two- or three-year payback periods; they are usually more concerned about initial costs. And when purchasing a home, they rarely consider energy efficiency (size, appearance and location are more important criteria).

Another impediment to greater efficiency is the fragmentation of the building trade. A commercial building project, for example, typically involves a series of handoffs: an architect designs the building but then hands it over to an engineer, who in turn specifies materials, systems and components, before passing responsibility to a contractor. Eventually the finished building is turned over to a maintenance and operations staff, who had virtually no say in the design process but probably know most about the building's day-to-day performance. In such a setting, initial construction costs are weighed much more heavily than long-term operating costs.

The system has to be improved. In the housing market, home buyers must learn to incorporate energy costs into their purchasing decisions, and home builders need to exploit energy efficiency as a profitable product that consumers value. On the commercial side, attention must be paid not only to the design of the building but to the way in which it functions as a whole. Because energy costs over a building's life span are comparable to the initial costs of construction, an increase of only a few percent in efficiency still translates into a considerable sum of money.

The potential for energy savings is probably greatest in the existing stock of commercial buildings, where energy costs may consume 30 percent of the operating budget. Indeed, energy efficiency in commercial buildings is growing rapidly, a trend that is attributable in part to a new generation of energy-services companies. Such companies offer expertise in the design, installation and long-term maintenance of sophisticated heating, ventilating and air-conditioning (HVAC) systems. Simply by fine-tuning existing HVAC and lighting systems, which together constitute more than two-thirds of a building's energy demands, an energy-services company can cut a building's energy costs and at the same time increase occupancy comfort. The savings (or avoided costs) can then finance more extensive retrofit projects, including major overhauls of equipment and operating systems. In the long run, everyone benefits: the building owner saves money, the energy-services company turns a profit and the drain on the global energy supply is lessened.

Because one of us (Bevington) is employed by an energy-services company — Johnson Controls — we have chosen to elaborate on two settings with which we are familiar: the Houston Intercontinental Airport in Texas and the Government Center complex in Boston, Mass.

At the Houston airport, Johnson Controls was asked to reduce heating and cooling costs for the passenger terminals without lowering comfort levels. Initiation of a maintenance routine (replacing filters and cleaning coils in the air-conditioning system) and scheduling the operation of individual chillers (air-conditioning units) according to the time of day and outdoor air temperature reduced energy costs at the airport by 20 percent, from $2 million a year to $1.6 million. These savings were achieved solely by optimizing the performance of existing systems; no capital investment was required.

Larger-scale retrofits generally demand substantial investments of capital. In most cases, the energy-services company provides the funding and guarantees a certain reduction in energy expenditures, but it expects to receive from 50 to 70 percent of savings accrued from the retrofit until payback is achieved. Government Center (a four-building, 2.25-million-square-foot complex) contracted a large-scale, multiple-stage retrofit involving windows, lights and the HVAC system in its two 22-story buildings. Replacing the windows was prohibitively expensive, so instead the existing windows were coated with reflective film to reduce solar radiation (solar heat) entering the building. The lighting fixtures, which account for 30 percent of the electricity in a commercial building, were retrofitted with reflectors (see Figure 3.3), thus reducing the number

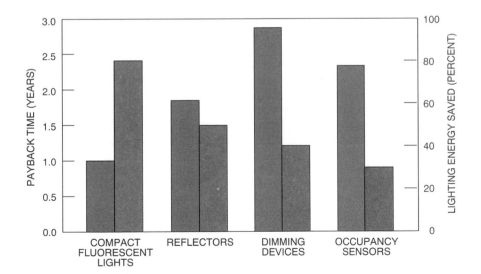

Figure 3.3 LIGHTING RETROFITS cut electricity costs (*pink*) and are relatively cheap, having payback periods of only one to three years (*gray*). Data from Johnson Controls show that compact 18-watt fluorescent bulbs (which provide the same light as a 75-watt incandescent bulb) can cut lighting costs by 80 percent. Reflectors redirect light, ena- bling half the bulbs to be removed, yet effectively maintain light levels. Dimming devices adjust the lights according to ambient sunlight. Occupancy sensors turn lights out when a room is unoccupied for more than, say, five minutes.

of bulbs per fixture by half but keeping the available light constant.

As part of the HVAC retrofit, maintenance crews cleaned and calibrated the air-handling system and had four chillers of various sizes installed. The advantage to installing chillers of differing size and cooling capacity is that an unnecessarily large chiller need not be run on a relatively temperate day, and so a significant amount of energy can be reduced in this way.

The heart of most building retrofits consists of a microprocessor-based integrated-control system, which monitors and operates the HVAC and lighting systems (see Figure 3.4). Sensors placed at strategic points within the building and throughout the HVAC system continuously feed information into the system. Parameters set for each region of the building, for example, maintain temperature and humidity, which allows for precise control of the interior climate. The system can selectively keep computer rooms at a lower temperature than, say, a storage area; it can also set ventilation rates and thus increase air flow to a smoking area or a cafeteria.

Johnson Controls spent more than $2.5 million retrofitting the government complex and installing the computerized management system. The result is a drop in the energy bill from an average of $6 million a year to $3.5 million. Moreover, for the 10-year life of the contract, both parties will benefit. Johnson recovers half the money saved in reduced utility costs (about $1.2 million a year); Massachusetts has the other half to spend elsewhere.

School systems across the U.S. have undertaken similar retrofits. Honeywell, a leading controls company, recently launched a lighting-modernization program in 12 Newark, N.J., schools. Installation of high-efficiency compact fluorescent light bulbs reduced maintenance requirements, doubled illumination levels and cut electricity use by 15 to 20 percent. Under the agreement, Honeywell has guaranteed that Newark will achieve savings of more than $200,000 a year, which enables the school system to pay for the retrofit from current operating funds. Although the payback period is relatively long—about five years—the buildings will last for another 50 years.

On a smaller scale, an integrated-controls system installed at the six-story First National Bank in Huntsville, Tex., has reduced yearly energy costs by $30,000 with a 2.5-year payback. In Overland Park, Kan., a similar installation has saved Humana Hospital $98,000 a year, an amount that will also be paid back in 2.5 years. In Houston the public library

system reduced its utility bills by 31 percent and so saved about $500,000 a year, with a payback period of three years. The capital made available as a result of these retrofits can meet more important needs such as hiring more teachers, training more physicians and buying more books.

The reason that energy-services companies are growing rapidly extends beyond their ability to reduce energy expenditures; it includes their willingness to provide long-term maintenance for a building. As microprocessor-based control technologies become more prevalent, businesses find it economically worthwhile to turn to outside contractors rather than to hire and train their own people to handle specialized equipment and sophisticated diagnostic tools.

Eventually self-diagnostic capabilities will be imbedded in the control systems themselves, reducing the need for large numbers of highly skilled operators, but as yet, such technologies are limited to small-scale, self-contained systems such as photocopying machines. When a photocopying machine malfunctions, for example, a display panel reveals the location of the malfunction and suggests step-by-step repair strategies. Someday "smart systems" will do the same for commercial buildings, reducing the need for highly skilled and scarce personnel and thus decreasing overhead costs.

Not all energy-reduction measures involve major retrofits. Indeed, in some cases, so-called passive strategies can produce excellent results. Recent studies have shown, for example, that exploiting a building's thermal storage capacity can reduce air-conditioning costs from 30 to 70 percent. The principle is simple: a building acts as a heat sink, which means that it stores heat and so (like a large cathedral) takes a long time to reach the same temperature as the surrounding air. Thus, in summer it is more efficient to let the inside temperatures gradually rise over the course of the afternoon; at the end of the day (when the building is no longer occupied), the cooling equipment is turned off, and temperatures inside are allowed to exceed the comfort range. During the night, when the outside air is cooler and electricity is cheaper, the equipment is turned on, and the building cooled. Precooling in this way reduces the demand for air-conditioning during the day when electricity rates are high and so saves on costs. Because the coolness of the building delays the buildup of heat, comfort is not compromised.

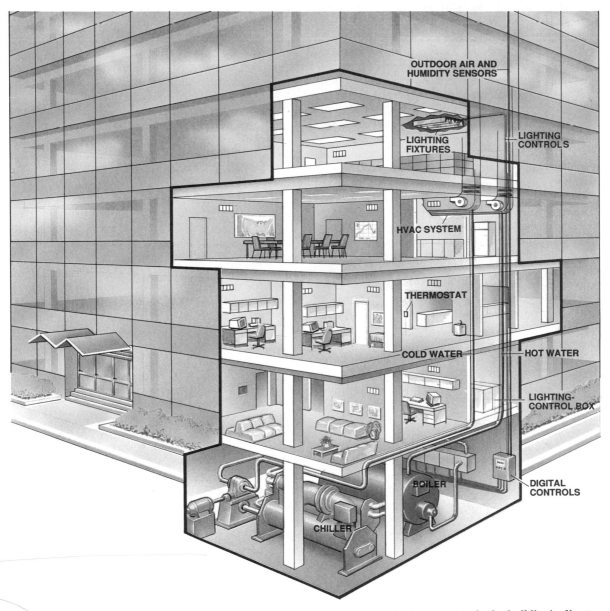

OUTDOOR AIR AND HUMIDITY SENSORS

LIGHTING FIXTURES

LIGHTING CONTROLS

HVAC SYSTEM

THERMOSTAT

COLD WATER

HOT WATER

LIGHTING CONTROL BOX

DIGITAL CONTROLS

BOILER

CHILLER

Figure 3.4 RETROFITTING BUILDINGS increases comfort and reduces energy costs. A large-scale retrofit generally calls for the replacement of lighting and HVAC systems and the installation of automated controls. Sensors placed at various points feed temperature and humidity data to the computer, which then controls the building's climate. In some cases, windows are coated with a reflective film, which lowers cooling costs by reducing the amount of infrared radiation entering the building.

The strategy contradicts contemporary thinking, which dictates that a building should be cooled only when it is occupied and only when absolutely necessary. Although not highly technical, the strategy requires a control system with a simple algorithm for forecasting the building's energy needs 24 hours in advance. The savings vary, depending on the local utility's rate structure, the building materials and the performance of the cooling system under different load conditions. Obviously, the better insulated the structure is, the more its thermal storage capacity can be exploited.

In Sweden thermal storage is utilized during the winter. All new office buildings are well insulated and store "free" heat—the heat generated by people, lights, office equipment and incoming sunlight. Whereas most office buildings in the U.S. have thin walls and cool quickly during a winter night, Sweden has well-insulated offices that store free heat during the day and overnight. By storing heat in the walls, floors and ceilings, the building becomes almost self-sufficient and thus does not need a central heating system.

Harnessing the power of the sun to illuminate the interior of buildings is another passive technology that is prevalent in Europe but has yet to be widely adopted in the U.S. The fact that one square foot of direct sunlight passing through a clear-glass window can illuminate 200 square feet of floor space (if it were evenly distributed) indicates the potential for illuminating a building's interior without electricity.

Ambient sunlight, however, is highly variable; indeed, the intensity with which it strikes a window may vary by a factor of 20, depending on the weather, time of day, season and so on. A window designed to meet overcast requirements admits too much light, glare and heat when it is sunny. Conversely, a window designed to screen peak intensities of sunlight admits too little during the rest of the year.

Other strategies can significantly reduce a building's need for electric lights. Interior shades or blinds equipped with photosensitive sensors can automatically adjust themselves according to the amount and direction of sunlight entering the window. Exterior shades serve a similar function by diffusing and directing the light before it strikes the window's outer face. Alternatively, windows can be equipped with reflectors and other optical devices that direct light deep into a room.

I n recent years engineers and manufacturers have made important advances in the design of the windows themselves. Most windows do not insulate well: they lose heat to the outdoors during the winter and admit excessive solar radiation into a building during the summer and thus account for 25 percent of all heating and cooling requirements in the U.S. Indeed, the energy drain from windows alone is equivalent to the amount of energy flowing through the Alaska pipeline each year.

Originally, windows were made from a single layer of glass. Although glass is transparent and can block the wind, it is almost as poor an insulator as metal and has a high emissivity: it readily absorbs and emits heat in the form of infrared radiation. On a cold day, a single-glazed window is nearly as cold as the outdoors, and its thermal resistance value is about R-1.

After World War II, factory-made double-glazed windows, known as thermopane or insulating-glass windows, were developed. They consist of two sheets of glass separated by a quarter-inch air space and have an R-2 rating, meaning they are twice as resistant to heat loss as are single-glazed windows. More recently, a low emissivity (low-E) coating has been applied to the inside surface of the glass facing the sealed air space. The low-E film reflects radiant heat back into the building rather than transmitting it outdoors. These windows are rated as R-3. Adding a second low-E film and substituting various gases for the air between the two sheets of glass achieves further savings. Argon, for example, which is nontoxic, improves the thermal characteristics of an R-3 window by 25 percent. Rated at R-4, an argon low-E window has twice the insulating value of a conventional double-glazed window.

It is worth noting that the average low-E coating plant costs $5 million to build, yet the windows it produces each day will save 10,000 barrels of oil, an amount equivalent to a 10,000-barrel-a-day offshore drilling platform, which costs 100 times more —$500 million—to build. Although all major U.S. window manufacturers now only make low-E windows, such windows are as yet unknown in Eastern Europe, the Soviet Union and China. Our advice to these countries (and to the World Bank) is to invest in coating plants; the initial investment is minimal, and the long-term financial and environmental benefits are substantial.

How far can the performance values of a window be pushed? Remarkable as it seems, the 1980's have given rise to a new generation of superwindows (see Figure 3.5). Creating a vacuum between the panes of a low-E window or filling the space with a transparent yet nonconducting material (such as aerogel, a high-technology insulator) can create resistance values that range from R-6 to as high as R-10. One prototype now being developed has actually performed better on a daily average in field tests than a highly insulated (R-19) wall. At night the superwindow loses a little more energy than an R-19 wall, but during the day even a minimal amount of sunlight is sufficient to turn the window into a net energy provider. In other words, superwindows

Figure 3.5 TWO WINDOW technologies are compared. Superwindows (*right pane at left*) that are rated as R-9 are as clear as low-E, R-4 double-glazed windows (*left pane*) but retain more heat during the winter, as shown by the infrared image. The double glazing produces a blue image, indicating heat loss, whereas the superwindow produces a yellow image, indicating it has retained heat almost as well as the R-50 wall.

gain heat, whereas the surrounding wall only prevents heat loss.

Superwindows may cost from 20 to 50 percent more than conventional windows, but their payback time (calculated in terms of energy savings) is from two to four years. They also have other advantages. They resist condensation, block ultraviolet rays that fade the color of furnishings and provide greater thermal comfort.

In warm climates, of course, heat gain is more of a problem than heat loss. Sunshine flooding through a window, as everyone knows, can raise indoor temperatures and so create tremendous cooling loads. Advanced window coatings now solve that problem. Such films selectively block solar radiation yet let visible light enter the room, thus reducing the need for artificial lighting and air-conditioning. Still under research and development are electronic and photosensitive windows; when the sunlight is too bright, they automatically reflect or absorb the light.

Great strides are being made in the housing market, where microprocessor-based technologies have given birth to what are commonly known as smart houses (see Figure 3.6). Such houses are equipped with automated-control systems, comparable to the integrated-control systems installed in commercial buildings. In fact, a number of manufacturers, utilities and trade associations—including the Electric Power Research Institute (EPRI) and the National Association of Home Builders—have recently formed a consortium to promote smart-house technology. Although the primary goal of smart houses is to provide their occupants with maximum comfort at an affordable price, many of the houses include energy-efficient technologies.

Southern California Edison Company, for example, has launched a House of the Future pilot project in cooperation with local builders. Each house is equipped with a touch-screen automation system that controls heating and air-conditioning, lighting and security. Simply by touching the screen, the owner can set temperature and humidity levels for any room in the house. The homes are also equipped with a range of energy-saving technologies such as compact fluorescent light bulbs, occupancy sensors (which automatically turn out the lights when a room is unoccupied for five to seven minutes) and energy-efficient appliances that can be programmed to run during the night when the demand for electricity is low.

Today a few builders are offering "optimum homes," which may be more expensive to build and thus have payback times of five to seven years but are much cheaper to operate over the long term (see Figure 3.7). Such homes typically have six inches of wall insulation (as opposed to the standard 3.5 inches) and an outer insulating sheath, giving the wall a total resistance value of R-24 (instead of the conventional R-11). The windows are low-E, double- or triple-glazed, and air flow in the homes is controlled by mechanical exhaust fans rather than by wind, allowing for more uniform and better indoor air quality.

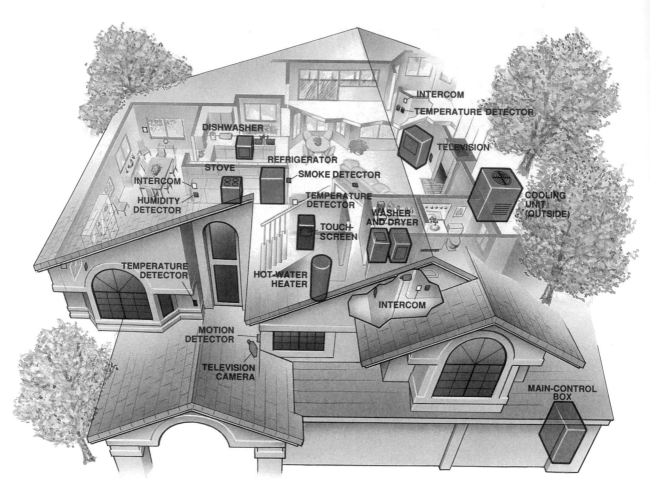

Figure 3.6 "SMART HOUSES" offer increased comfort at a reasonable price and are far more energy efficient than the average home. At the heart of each house is an automated-control box (*purple*) that monitors heating, air-conditioning, lighting and security systems. In more advanced homes the owner can adjust temperature and humidity and turn appliances on and off by touching a wall-mounted screen (*red*). Passive measures, including well-insulated walls, roofs that reflect solar radiation and shade trees (particularly on the south and west side of the building) that mitigate the sun's heat, also save energy.

Another innovation is a hot-water heater that doubles as a furnace. By diverting hot water to small heating units, builders of superinsulated homes can save $1,000 to $2,500 (the amount normally spent on a central furnace and ventilating system) and invest in more insulation. Even on a cold Chicago night, when the temperature may fall to −25 degrees Celsius (−13 degrees Fahrenheit), a superinsulated home is kept warm by its hot-water heater. Although such homes are still rare in the U.S., they are now standard in Scandinavia, where they are prefabricated in factories and shipped in truck-size pieces to their final destination.

Optimum and superinsulated homes also present an unusual marketing opportunity. They are so energy efficient that a Chicago builder, the Bigelow Group, guarantees their low operating costs. If the annual heating bill of a town house exceeds $100 ($200 for a single-family home), the builder will pay the difference. In addition, each year Bigelow holds a contest to see who has the lowest heating bill; last year's winner had an annual heating bill of only $24, not bad given the severity of Chicago winters.

Houses, like commercial buildings, are ideal subjects for passive conservation measures. One of the simplest, oldest and yet most effective strategies for

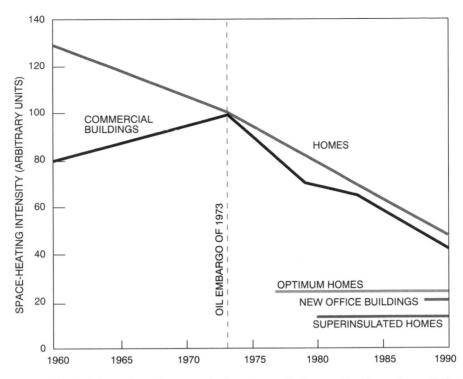

Figure 3.7 SPACE-HEATING intensity, the amount of heat per unit floor area needed for a comfortable inside temperature, has dropped by almost 50 percent in new U.S. buildings. The decline has resulted from efficiency improvements spurred by the oil embargo of 1973. Leading the revolution are "optimum homes" that are insulated (but still have forced-air heat), offices that have state-of-the-art windows and building shells, and superinsulated homes that need no central heating.

reducing a home's heat gain during the summer is to mitigate the sun by planting shrubbery and painting the house a reflective color. On a hot (85-degree-F.) afternoon, an unshaded tract home draws from two to five kilowatts of power for cooling, whereas a well-shaded house, painted white, consumes one kilowatt less during the same period.

In addition, most large U.S. cities have become summer "heat islands." In Los Angeles, downtown temperatures are three degrees C. higher than they were in 1940 and are increasing by .5 degree C. every 10 years because of the spread of asphalt across the city. Not only is asphalt dark and thus a heat absorber, but it replaces trees and so reduces cooling by evapotranspiration. Passive strategies such as planting shade trees and whitening buildings and asphalt can reverse the Los Angeles heat-island effect, which now costs an extra $100,000 and adds 1,000 tons of CO_2 to the atmosphere every hour. These measures thus exert a positive effect on the environment by preventing a significant

amount of carbon dioxide from reaching the atmosphere.

It is readily apparent that the opportunity for reducing energy expenditures in buildings has never been greater nor has the need been more pressing. Mounting environmental concerns and industrial competitiveness have provided powerful incentives to embrace the wealth of new technologies now available. Yet the remarkable rate of progress achieved from 1973 to 1986 has slowed. How can interest in efficiency be rekindled? What measures are most likely to force the curve upward once again?

The most practical strategy—at least in the U.S.—is to impose national building and appliance standards that are supplemented with incentives to beat those standards. At present, the American Society of Heating, Refrigerating and Air-Conditioning Engineers (ASHRAE) regularly updates voluntary standards, keeping them at about a three-year payback time. Although the ASHRAE Series-90 standards are adopted by most states, they are poorly enforced. The result is that most new homes or buildings have the poor efficiency associated with payback periods of only two to three years. Ironically, the federal government has adopted standards for its own buildings that are even tighter than the ASHRAE series, but it seldom enforces them and seldom installs high-efficiency lights or windows or other cost-effective innovations. Nor does it stress tune-up and maintenance.

Some states, notably those on the West Coast, have tighter standards that are conscientiously enforced and therefore are willing to accept slightly longer paybacks of three to four years. Similar standards, called norms, are commonplace in Western Europe.

Such standards, however, apply only to new buildings. Equally important are standards for commercial retrofits and rental properties. Some cities in California have recently addressed this problem by enacting residential conservation ordinances (RECO's) and commercial-building conservation ordinances (CECO's), which stipulate that a building must be upgraded to minimum standards (those that have a three-year payback) before its title can be transferred. A special conservation ordinance is now being considered just for rental property. Because rental units are generally so inefficient, significant savings can be achieved even when improvements that have a two-year payback are made.

An effective strategy for new buildings is now under consideration in Massachusetts. There the state legislature is currently debating a bill (House Bill 5277) that sets revenue-neutral "feebates" for commercial buildings 50,000 square feet or larger. Buildings that intend to use more electricity per square foot than average would be charged a stiff utility hookup fee. Buildings that are designed to use less electricity than average get a rebate. The system is revenue neutral because virtually all the fees collected from the extravagant buildings are rebated to those that consume relatively little electricity. Only a small portion goes to administrative costs and to the utility for promoting energy-saving measures.

For almost a decade, Fannie Mae (the Federal National Mortgage Association), Freddie Mac (the Federal Home Loan Mortgage Association) and the Veterans Administration have approved bigger loans (and lower-interest rates) for homes that are certified as energy efficient. Yet only one in 10,000 U.S. home buyers takes advantage of these programs; clearly, better publicity is needed, and the programs should be expanded.

The situation may change now that a number of utilities and even some home builders are offering Home Energy Rating Systems (HERS), which certify that a home meets a specified level of energy efficiency or that it contains specific features such as a gas dryer (instead of an electric one). The HERS function on two levels: they enable a potential buyer to make an intelligent decision about the actual efficiency of a house (and what its utility bills are likely to be), and they provide builders with an incentive to construct affordable, energy-efficient homes (it is now apparent that HERS are a selling point in the housing market).

We also believe that research devoted to energy efficiency for buildings must be intensified. The building sector is itself hamstrung; dominated by thousands of small businesses, it is far too fragmented to support the type of research carried out by either the transportation or industrial sector. The burden then falls on the government and utilities. Yet together they spend only about $200 million a year, or .1 percent of the country's annual $200-billion utility bill for buildings. Such an amount is clearly inadequate. Even mature industries — such as steel, automobile and chemical — allocate about 1 percent of their revenues to research and product development, an amount 10 times greater than that spent by the building sector.

Sweden, by comparison, has formed a National Council for Building Research, with an annual budget of $1 billion (scaled to the U.S. economy, which is 30 times larger than that of Sweden). Not surprisingly, the results of such a program have been impressive. Today Sweden leads the world in having the largest percentage of energy-efficient buildings and exports the greatest fraction of building technology around the world. In fact, when scaled to the U.S., its building sector achieves a $60-billion trade surplus per year, whereas the U.S. building sector accounts for an annual trade deficit of $6 billion.

The future, although by no means secure, appears promising. The wealth of available technologies has created a truly vast potential for energy savings in all types of buildings. Commercial buildings that today consume annually an average of 15 kilowatt-hours a square foot will be driven from the market. In their place will rise buildings that consume less than five kilowatt-hours a square foot. Getting from here to there is not just a matter of time; it requires our unwavering commitment.

Energy for Industry

Industrial processes consume two-fifths of the developed world's energy. Efficiency improvements have steadily cut that share and promise to continue.

. . .

Marc H. Ross and Daniel Steinmeyer

A fortunate paradox has overtaken industrial production during the past 20 years: output has risen substantially, but total energy consumption has gone down. Companies engineered this apparent contradiction by making industrial processes more productive and by investing in conservation as energy prices increased (see Figure 4.1). Consumers have also helped reduce industrial energy consumption by favoring products that require less energy in their manufacture and use.

From 1971 to 1986 the energy intensity of industrial processes in the U.S.—the amount of energy required to produce a ton of steel or a kilogram of polyethylene—has declined by between 1.5 and 2 percent a year (direct fuel intensity fell at a 3 percent rate while electrical intensity remained roughly constant). These efficiency improvements, together with changes in the mix of products people buy, have led to a decline of 1 percent a year in total energy use despite a 2 percent annual growth in manufactured output. (The fragmentary data available for more recent years suggest that improvement is continuing but at a much slower pace.)

These savings are important because industrial processes account for about 40 percent of the energy used in the developed world (see Figure 4.2). More than half of that goes to convert ores and feedstocks to basic commodities, such as steel and gasoline; the rest serves agriculture, mining, con-

struction and the manufacture of intermediate and finished goods (see Figure 4.3).

The reason industry has been able to reduce energy use so consistently is simple: virtually every current manufacturing process uses far more energy than the minimum required by thermodynamic laws. These laws specify the minimum amount of energy required to do physical work—to lift a mass from a low place to a higher one, to separate a mixture into its components or to drive a chemical reaction to a higher-energy state.

As carried out in even the most efficient plants, however, basic operations, such as the separation of oxygen from air and the production of ethylene from petrochemical feedstocks, expend between four and six times the thermodynamic minimum. Furthermore, many industrial processes that now consume vast amounts of energy, such as the production of metal parts in particular shapes or the attachment of one part to another, theoretically require no energy at all.

This huge margin for improvement has made possible the reductions in energy use achieved so far and will continue to do so in the future. Continuing reductions in energy consumption will go a long way toward meeting the goal, proposed by researchers concerned with global warming, of a 20 percent reduction in carbon dioxide emissions in the next two decades.

Reducing energy consumption begins with some basic physical laws. First, strictly speaking, energy is neither consumed nor created. What is consumed is the capacity of energy in a particular form to do work. A coal-fired electric-power plant, for example, transforms the chemical energy in a lump of coal into the thermal energy of superheated steam that drives a turbine and generates electricity. In this case, about 35 percent of the energy in the coal is converted to electricity; the rest remains in the combustion gases going up the boiler's stack and in the steam leaving the turbine, which have little capacity to do work.

Any practical transformation that converts energy from one form to another loses some of its capacity to do work. Consequently, an even better solution than using energy efficiently is not using it at all. As a result, some activities that apparently have little relation to energy, such as quality control or the recycling of scrap, may have an enormous impact on energy consumption. Recycling bypasses the most energy-intensive steps of manufacturing—the conversion of ores and feedstocks into basic materials. Effective quality controls save all the energy that otherwise would have been used to produce defective products.

In practice, companies reduce energy consumption—the conversion of energy from useful to nonuseful forms—by optimizing the cost of existing processes, by introducing process refinements and by making breakthroughs that lead to entirely new methods of manufacture. The first of these, cost optimization, is straightforward: as the cost of energy rises, companies can substitute equipment (thicker insulation, larger heat exchangers and other energy-saving devices) for energy, thus reducing consumption.

The second—process refinement—operates even in the absence of rising energy prices. The technical improvements reduce the total costs of an industrial process by 1 to 2 percent a year on average. The refinements typically reduce energy consumption as well as labor and materials costs. Because they usually increase production capacity, technical refinements reduce capital costs instead of increasing them as optimization does. (They also generally reduce the production of wastes—anything that improves the productivity or efficiency as a process usually yields a cleaner operation.) Examples range from automated controls to improved catalysts for chemical reactions.

Finally, scientific and technological breakthroughs can enable relatively rapid and profound reductions in industrial energy requirements. The float-glass process, for example, first adopted in the mid-1960's, casts sheets of plate glass on a smooth layer of molten tin. It eliminates entirely the energy formerly required to grind and polish the glass after solidification.

Similarly, the basic-oxygen conversion process for steelmaking, adopted in the 1960's, consumes only about half the energy used by the open-hearth methods that preceded it. Oxygen blown through the molten metal burns most of the carbon that is present—the precise level can be adjusted as desired—and the heat generated inside the metal by the burning supplies energy to remove other impurities in the form of slag. (Open-hearth furnaces were heated entirely from the outside.) Additional process refinements, now used in Japan, capture the chemical energy in the gases emitted from the converted and so cut outside energy consumption to near zero.

These disparate examples of falling energy use underline the diverse nature of industrial activity. Unlike the housing or transportation sectors, where a few basic solutions may be widely applied to reduce overall energy consumption dramatically, manufacturing requires a case-by-case approach. A few energy-conservation devices—variable-speed motors for pumps (instead of constant-speed motors and flow-reducing valves), heat exchangers (instead of coolers for one stream and heaters for another) or automated process controls—barely begin to characterize the variety of opportunities for improving thousands of separate processes.

To optimize the energy consumption of an industrial process, one must break the process down into a series of steps and compare the theoretical amount of energy required for each step to the energy actu-

Figure 4.1 STEEL FURNACE exemplifies the energy consumed in converting raw materials to commodity goods such as metals, glass, plastics and paper. New processes (such as direct steelmaking or making steel from recycled scrap in an electric-arc furnace) may eventually cut industrial energy to a small fraction of its current level.

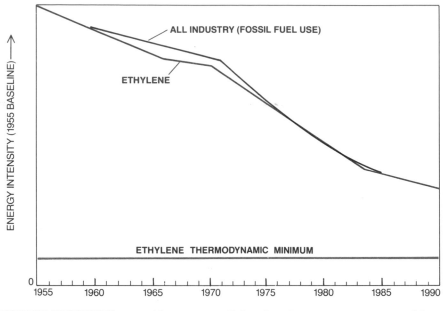

Figure 4.2 INDUSTRIAL PROCESSES account for approximately 37 percent of all the energy expended in the U.S. Industries that convert raw materials are the largest consumers.

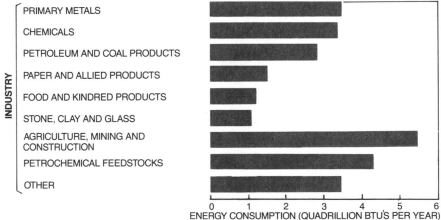

Figure 4.3 ENERGY INTENSITY (the amount of energy required per unit of output) declines as a result of competition. For example, engineering contractors sell facilities that make ethylene, a major chemical feedstock, on the basis of product yield and energy required per pound. Between 1955 and 1990 energy requirements dropped by nearly two-thirds, although consumption is still above the minimum required by thermodynamic laws. Much of the credit goes to low-cost thermodynamic analyses made possible by computers. The amount of fuel consumed by U.S. industry overall per unit of output has declined by more than 50 percent during the past 30 years.

ally expended. It is not very helpful to know, for example, that a distillation column that separates ethylene from ethane consumes five times the theoretical minimum energy; what is helpful is knowing which specific parts of the process are responsible for most of the loss in the input energy's capacity to do work.

If 60 percent of the "lost work" results from temperature differences between the process stream and heating or cooling streams, which add heat to the process stream or subtract heat from it, then it becomes clear that energy can be conserved by reducing temperature differences (see Figure 4.4). That could be accomplished, for example, by employing larger heat exchangers and redesigning refrigeration equipment so that it produces temperatures closer to those at which the process operates.

This type of efficiency improvement relies on the fact that different tasks can be carried out by different sources of energy. Considerable savings can be achieved by making good matches: steam that is no longer capable of driving a turbine efficiently may

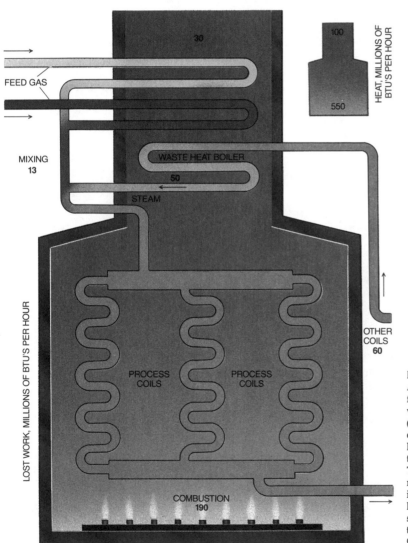

Figure 4.4 A THERMODYNAMIC ANALYSIS gives designers "new eyes" for reducing energy use. A simplistic view of a chemical reforming furnace (inset) shows only the combustion energy going in (550 million Btu's per hour) and the amount of heat leaving the stack (100 million Btu's per hour). The potential to save energy simply by recycling heat from the furnace exhaust is limited. A more complex analysis, however, reveals increases in potential savings by focusing on "lost work"— the ability of the heat to perform useful tasks.

still be perfectly suited to heating a chemical feed-stock to the temperature required for a reaction or to drying paper in a papermaking plant.

The decision whether to trade capital for energy by installing larger heat exchangers or other more efficient equipment is largely a matter of economics. Companies seek a balance between the yearly costs of the energy required to drive their manufacturing processes and the annual costs (interest on and repayment of principal) of capital investment to reduce energy consumption. The cost of piping, heat exchangers, insulation and similar items increases smoothly as a function of size, as does their combined contribution to energy efficiency.

Such optimizations are not always easy to make in practice. The yearly cost of capital for many facilities far outweighs the cost of the energy used to operate them. (The chemical industry, for example, spent roughly three times as much on capital equipment in 1989 as it did on energy.) As a result,

companies can rarely economically discard old equipment simply because rising energy prices have made it less efficient than would be desirable.

A fourfold increase in energy prices (such as that caused by the 1973 oil shock), for example, leads to a 70 percent decrease in the optimum energy used to pump fluid through a given length of pipeline. But reducing the energy consumed in a piping system by that amount would lead only to a 13 percent drop in total costs (see Figure 4.5). The energy savings may easily justify conservation measures in a new facility, but it will not necessarily justify the cost of replacing equipment in existing plants.

Other factors also delay the implementation of energy-saving technology. Engineers may simply be unaware of the appropriate energy-conserving action for their particular industrial process. And even if they do know what to do, they may have trouble convincing management to let them proceed. Managers may be unsure whether it makes more business sense to retrofit facilities with energy-efficient devices, to build new facilities and shut down old

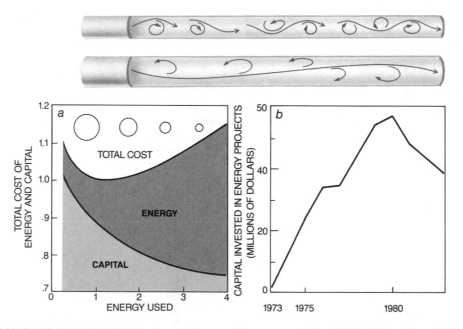

Figure 4.5 ECONOMIC OPTIMIZATION of energy use leads to dramatic decreases in energy consumption but only to a small reduction in overall cost. A properly sized piping system, for example, typically costs seven times as much as the energy required to pump fluid through it. As the diameter of the pipe decreases, its capital cost falls, but its energy cost rises (a). Installing a larger-diameter pipe to reduce frictional losses might be justified in a new plant but not in an older one. This trade-off helps to explain why, although energy prices skyrocketed in 1973, many companies did not make major energy-saving investments until years later (b).

ones, or simply to wait and see if energy prices come back down (as they did in the early 1980's).

The choice is a particularly difficult one for a marginal firm, where poor choices could lead not to reduced profits but to bankruptcy. The common response of most businesses is to impose a short time horizon on cost-cutting investments such as energy conservation so that top managers can devote their attention to the development of new markets and financial arrangements. Companies in the U.S. typically require that an investment pay for itself within about two to four years.

Economic trade-offs between energy use and capital equipment provide incentives for reducing energy consumption only when energy prices are rising faster than the cost of capital equipment. Process refinement is a much more consistent force for improving efficiency because it does not balance energy against capital—instead it can reduce energy consumption and capital costs at the same time.

In most manufacturing processes, for each doubling of cumulative production, total processing costs—including energy—drop by about 20 percent. Often energy savings are merely a by-product of changes made to improve quality or increase productivity. This relation has been found to hold for processes as disparate as steelmaking and the production of polyethylene. The learning curve does not depend on increasing energy costs; indeed, substantial energy improvements were made in these and other processes during the 1950's and 1960's, when energy prices were low and in many cases even declining.

Improvements in the production of ammonia furnish a typical example of process refinement. In an ammonia plant, streams of hydrogen and nitrogen gas pass over a catalyst at high pressure, causing a portion of them to combine. Unreacted gases are recycled. A new geometric layout of catalyst surfaces, developed by the M. W. Kellogg Company, an engineering design firm in Houston, Tex., increases the contact between the reactants and the catalyst, thereby converting a higher percentage to ammonia on each pass and increasing the production rate by 6 percent. The layout simultaneously reduces the pressure needed to force the reactant through the catalyst bed, reducing energy consumption by about 5 percent.

The introduction of rubber-coated air bags to cushion the dies used in stamping presses is an example of process refinement in more conventional manufacturing. In standard practice the lower dies in these machines, which shape sheet-metal parts such as automobile-body panels, rest on compressed-air pistons, which tend to leak. The air bags do not leak, and so they consume much less air than the pistons. The bags' developer, Smedberg Machine, has found that retrofitting presses reduces plantwide compressed-air requirements by 50 percent and overall electricity consumption by 25 percent. Similar savings are realized in avoided maintenance, and the air bags significantly reduce the risk that the production line will be shut down by the failure of a single leaky piston.

The widespread adoption of automated process controls shows even more strongly how energy savings come as a by-product of meeting other goals such as greater safety and reduced staffing. Distillation columns, which separate materials on the basis of their different boiling points and account for 5 percent of all industrial energy consumption, were once run by so-called feedback controls (see Figure 4.6). Operators adjusted the flows of steam and feedstock depending on variations in the composition of the products coming out of the column. Typically they used far more steam than the minimum required so as to ensure that the column could tolerate swings in feed composition or temperature without catastrophic failure.

Today sensors in a column and its feed lines provide enough data for feed-forward operation—adjusting steam and product flows according to heat- and mass-transfer equations so that the column's final product will have the desired composition. Feed-forward operation allows columns to be operated at higher feed rates; a fully automated plant thus costs less to build than a plant with the same production capacity but controlled only by feedback. More relevant, energy consumption drops by between 5 and 15 percent because the feed-forward controls can adjust for changes in feed while using much less excess steam.

In the paper industry, another major energy consumer, automated controls can optimize the combination of heat and chemicals required to produce high-quality pulp, and they can also schedule the timing of operations to reduce peak power requirements, thus cutting energy costs further. In one mill, sensors and controls reduced variations in pulp quality by 31 percent while simultaneously reducing steam use by 19 percent.

Not all process refinements are quite so dramatic,

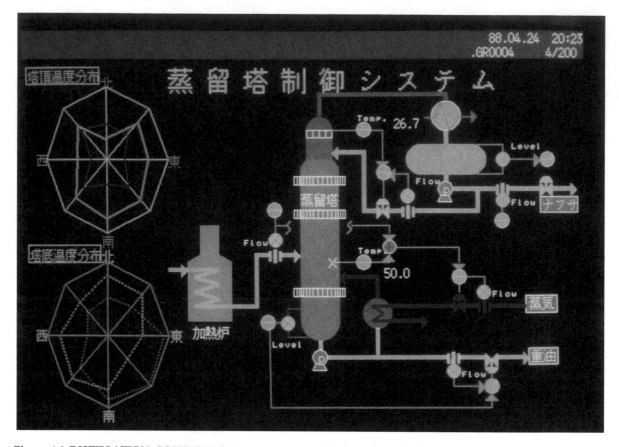

Figure 4.6 DISTILLATION COLUMN (*pink*) appears on the screen of an automated process-control system. Circular patterns (*left*) indicate the temperature at various points in the column. Piping diagram (*right*) shows the status of material fed into the column, extracted from it and recycled through it. Such process controls can reduce energy requirements by 10 percent or more; they may also make feasible new manufacturing methods that could not be carried out under manual control.

but a chain of small improvements accumulates over time to yield substantial dividends. For example, the yearly energy improvements in the process that produces ethylene from ethane have averaged 3 percent since 1960. Gains came from many different sources: more efficient motors, pumps and compressors; chemical processing sequences designed to exchange energy between hot and cold streams; improvements in process controls to produce a more consistent product; and selective furnaces to increase the yield of ethylene relative to other compounds. But together these "minor" savings add up to a 60 percent decline in the amount of energy required to produce a pound of ethylene from ethane.

The example of ethylene is particularly important because the chemical industry accounts for more than 15 percent of industrial consumption (more than any other), and ethylene production is the single largest user of energy in the chemical industry. It and its coproducts are also the feedstocks for the synthesis of most other organic chemicals.

Yet despite the large efficiency gains made thus far, substantial room for improvement remains. Even today the amount of energy used in making ethylene from ethane is about four times the minimum required by the laws of thermodynamics.

Even when carried out in the most efficient manner, some processes are unavoidably energy intensive. Substantial overall energy savings are occurring nonetheless because the share of industrial activities that consume less energy per unit of economic output is increasing. Shifts in the composition of production have contributed about 15 of the 35 percent drop in energy per unit of economic output since 1971.

The first steps of manufacturing—the initial conversion of raw material—are the most energy intensive. For example, an aluminum smelter spends $1.20 on energy for every dollar spent on wages and capital; a manufacturer of inorganic chemicals, such as oxygen or chlorine, spends 25 cents. But a maker of frozen foods spends only five cents on energy for every dollar spent on wages and capital, and a computer maker only 1.5 cents.

As the economies of the industrialized world have grown beyond supplying just the basic needs of their citizens, products that take relatively little energy to manufacture compared with their value (such as computers, medical instruments and pharmaceuticals) have accounted for an increasing proportion of economic activity. People are buying more electronic equipment and fewer major appliances—there is a limit, after all, to the number of refrigerators or automobiles that a household can absorb. Since the 1950's the relative contribution of basic materials to the U.S. gross national product has dropped by nearly 40 percent.

In addition to making new kinds of products that use less material and energy, manufacturers are making their old products from less material overall. Thin sheets of high-strength steel alloy substitute for thicker sheets of conventional steel in automobile bodies. Hair-thin glass fibers replace bundles of copper wire in telephone lines. Even sheets of paper are made thinner than they once were. The relative consumption of all major materials—glass, steel, cement, paper, fertilizer, chlorine—is declining. Even the use of plastics may have crested: consumption in the U.S. dropped 2 percent from 1988 to 1989.

Whereas the "dematerialization" of manufactured goods results in products that consume less of each basic material, recycling largely avoids the energy consumption needed to produce these materials. Typically it takes half as much energy to make recycled materials as it does to make virgin ones (see Figure 4.7). Only about 20 percent of all the paper, plastic, glass and metal goods in the U.S. are now made from recycled materials, out of

roughly 50 percent that might be. The potential energy savings are staggering.

Companies usually recycle almost all the scrap generated during the manufacturing process, but once an item has passed through consumers' hands the rates drop sharply. Only about 40 percent of the inputs to steelmaking consist of material recycled from sources outside the mill. Similarly, only about a quarter of the fiber inputs to paper mills are recycled material. And only a tiny fraction of the inputs to plastics manufacture consist of used goods.

It is difficult to find markets strong enough to absorb the potential supply of recycled materials: there is only so much demand for cellulose insulation from old newspapers, synthetic lumber from recycled plastics or reinforcing bars from recycled steel. The challenge is to create recycling systems that produce high-value products from scrap—such as automotive sheet metal made from automotive sheet metal or plastic bottles made from plastic bottles. So far the only major recycling effort of this kind is the aluminum beverage can, a high-technology product that contains more than 50 percent recycled material.

Economic optimization of energy use, gradual process refinements and shifts away from the production of energy-intensive goods can all lower the amount of energy consumed for a given level of economic output. Nevertheless, they are not enough.

In the long run, technological breakthroughs are the most dramatic factor in cutting back industrial energy consumption. Refinements of existing industrial processes would have run into their natural limits long ago were it not for radical innovations (see Figure 4.8). Limits on the wood available for the masts of ships, for example, were made irrelevant by the development of steel ships and alternative means of propulsion. Concern about the limited power available from waterwheels led to the development of the steam engine.

A more recent example shows how radical innovation and process refinements interact. Polyethylene, which accounts for a third of the 40 billion pounds of plastic produced in the U.S. last year, began its commercial life in the early 1940's. It was produced under very high pressure (12,000 atmospheres), but by the mid-1970's the energy required to produce a pound of polyethylene had been cut in half. Meanwhile, in the 1950's, two European chemists made some fundamental discoveries that led to a radically new production process based on

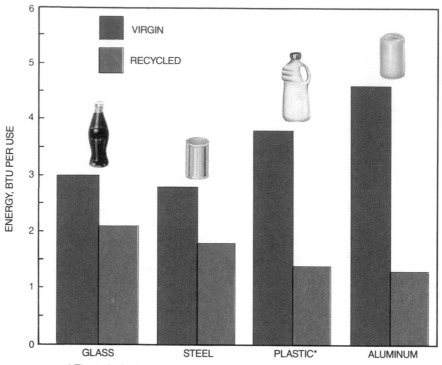

Figure 4.7 RECYCLING discarded containers consumes far more energy than refilling them; nevertheless, it still provides a comfortable margin over making containers from raw materials. Aluminum, which has the largest differential, leads in recycling, but plastic, with the next largest differential, is far outpaced by glass, which has the smallest.

solvents; this new learning curve led to Union Carbide's development, in the 1970's, of the low-pressure gas-phase process for making polyethylene. The new process is simpler and much safer; it now demands only a quarter the energy and half the capital of even the improved high-pressure process. Moreover, it even yields a stronger polymer.

Some breakthroughs, such as the low-pressure polyethylene process, are based on a combination of existing technology and new scientific discovery. Current worldwide efforts to develop a new technology for making steel directly from ore and coal are more clearly derived from existing technology —in this case, the basic-oxygen converter currently responsible for the bulk of global steel production.

In the proposed direct steelmaking process, powdered ore, coal, oxygen and flux are blown into the converter, which contains a bath of molten iron. The iron oxides are reduced to iron and the impurities driven off in the form of slag. If this technique can be made practical, one reactor, capable of continuous operation, would replace four batch processes (agglomeration of ore into pellets, conversion of coal into coke, reduction of ore in blast furnaces and basic-oxygen steelmaking). Capital costs would decline dramatically, as would the environmental damage now done by coke ovens. Energy savings of about 25 percent would also accrue.

Other industrial breakthroughs stem from the discovery of wholly new scientific phenomena. Quantum physics and the invention of the transistor, for example, led to the development of microprocessors. These minuscule computers can convert complex responses from physical sensors into accurate measurements of chemical composition, pressure, size and other quantities. They provide the basis for automated control of industrial

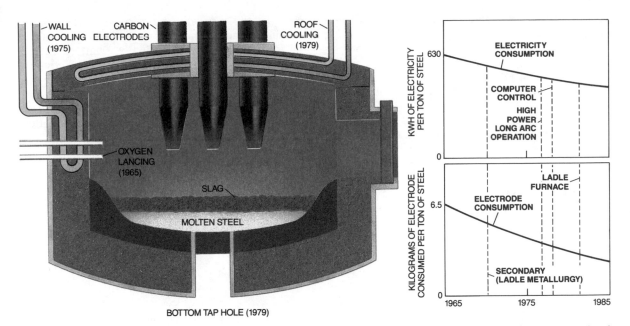

Figure 4.8 GRADUAL PROCESS REFINEMENT lowers overall production costs, including energy consumption. Electric-arc furnaces for steelmaking (*left*) have achieved total energy gains of greater than 30 percent since 1965, thanks to a series of incremental improvements in their various components. Two measures of furnace perform-ance are electricity used per ton of steel and pounds of electrode consumed per ton of steel (*right*). Over time, re-finements produce smaller savings, and so long-term re-ductions in energy consumption depend on breakthroughs that lead to entirely new industrial processes.

processes. And the automated controls, of course, have ushered in any number of new industrial techniques—including the use of robots in areas as diverse as automobile assembly and medical analy-sis.

Sometimes scientific discovery and commercial application are so tightly linked that they appear to proceed almost side by side. Examples include the interaction between polymer chemistry and devel-opment of synthetic fibers, such as nylon, ultra-long-chain polyethylene and kevlar, and the inter-action between solid state physics and micro-electronics.

Other discoveries, of course, have had minimal impact on commercial invention. The effects of groundbreaking discoveries in high-energy physics and nuclear science, for example, have been con-fined largely to the military.

Nevertheless, any major discovery offers a set of possibilities that did not exist before. New under-standing of molecular biology has a particularly rich potential. Modified plants that fix their own nitro-gen from the air could eliminate the need for nitro-gen fertilizers for crops such as corn and wheat. Although the proper manipulation of the 50 or so genes involved may take decades, the energy and environmental impacts are enormous. The manu-facture of these fertilizers currently consumes 2 per-cent of all industrial energy, and their use is be-lieved to be responsible for the major share of human emissions of nitrous oxide, a gas implicated in the greenhouse effect. Other organisms and agri-cultural plants will probably be engineered to pro-duce chemical products at costs lower than at present, with no input of petroleum feedstock.

To succeed, technological developments must be compatible with growing concerns about environ-mental quality. As a burgeoning world population raises its per capita consumption, these concerns are becoming increasingly important. Companies that do not recognize such social concerns face both di-rect and indirect costs, from lawsuits, project delays and the concerns of their own employees.

Although the reduction of industrial energy use depends on a myriad of case-by-case changes and optimizations, some general policy principles emerge. First, improvements must rest on a strong educational foundation. Good science teaching in elementary and junior high schools is fundamental to the future supply of scientists and engineers — and to the understanding of their contribution by nonscientists.

Equally important is the willingness of industry to use these trained people to maximize energy efficiency. Companies must have adequate profit margins from which to invest in research and development and in new technology. There must be strong competition between companies within each industry. At least as important is competition between industries that supply the same basic need (as, for example, the competition among makers of steel, aluminum and engineering plastics to supply material for automobile parts). Finally, plant operators, equipment suppliers, engineering firms and government regulatory bodies must all have long time horizons. The financial policies that have pushed U.S. firms to think in terms of fiscal quarters instead of decades must be changed. Favorable depreciation treatment of productivity investments would be a start toward redressing the balance.

The public focus on energy that arose from the oil shocks of the 1970's has long since dissipated. A renewed consensus is needed to spur the kinds of gains that will be necessary in coming decades.

Energy for Motor Vehicles

*They consume a growing share of the world's oil supply
and are also major polluters. Efficient designs, alternative fuels
and rational transportation systems can help solve the problem.*

. . .

Deborah L. Bleviss and Peter Walzer

About half of the world's oil is consumed by a fleet of 500 million road vehicles whose growth has consistently outpaced that of the human population. These vehicles account for most of the energy used in the carriage of people and freight. Since 1970 the fleet's annual increase has averaged 4.7 percent for cars and 5.1 percent for buses and trucks. If the trend countries, a billion vehicles will ply the world's roads by the year 2030.

Such rapid growth poses many problems (see Figure 5.1). In the long run, it causes oil consumption to rise faster than oil production, squeezing supplies. Indeed, the ongoing rise in oil prices appears to be putting an end to the buyer's market that emerged in the early 1980's (when an economic slowdown and conservation efforts temporarily reduced world demand). As prices rise, political power can be expected to shift back to the oil-exporting countries of the Middle East, leading perhaps to another wave of politically and economically disrupting oil crises reminiscent of those of 1974 and 1979 (see Figure 5.2).

Equally worrisome is the effect the burgeoning number of vehicles will have on regional and global environments. Regional air pollution threatens health, and much of it can be traced to the emissions of motor vehicles. The three major emissions are carbon monoxide (which displaces oxygen in the blood), nitrogen oxides (which react with water to form nitric acid) and hydrocarbons (which react with nitrogen oxides in the presence of sunlight to form ozone, a lung irritant). In the industrialized countries of the West and Japan—member states in the Organization for Economic Cooperation and Development (OECD)—motor vehicles emit nearly half of the nitrogen oxides, two thirds of the carbon monoxide and nearly half of the hydrocarbons. In developing countries, where environmental controls are lax, inefficient vehicles also contribute significantly to air pollution, even though fewer of them are on the road.

Vehicular emissions aggravate global environmental problems. A tank of gasoline produces up to 400 pounds of carbon dioxide, a major greenhouse gas implicated in global warming. Although the world's motor vehicles now produce only 14 percent of all the carbon dioxide derived from fossil fuels, the vehicular contribution in industrialized countries is higher, reaching a peak of 24 percent in the U.S., where per capita ownership of motor vehicles is highest in the world. Chlorofluorocarbon refrigerants, another kind of greenhouse gas, are now

being phased out of automotive air-conditioning systems.

Today most road vehicles either are manufactured in the OECD countries or are based on designs that originated in them. Hence, policies to conserve the world's energy and reduce vehicular emissions must concentrate on the industrialized nations (see Figure 5.3). These policies will generally go hand in hand: when other factors are held constant, a savings in fuel produces a reduction in tail pipe emissions. The relation is most obvious when administrative steps are taken to limit the number of miles vehicles may travel. Such rationing, however, is both unpopular and costly: economic growth depends on transport.

Vastly more attractive alternatives to rationing do exist. One can increase the efficiency with which vehicles consume gasoline and diesel, introduce cleaner fuels, redesign the road system, promote mass transit and change patterns of settlement so that people live closer to their work. Each of these solutions has its drawbacks, but they can be overcome provided enough political will is mustered.

So far most oil-saving steps have involved refinements in the design of gasoline and diesel cars. During the past 15 years the average car's consumption of fuel has fallen by a quarter in West Germany; in the U.S., where the initial level was higher, fuel consumption has fallen by half. The average car in the OECD countries now achieves nearly 30 miles per gallon (mpg). At the same time, emissions of major urban pollutants have dropped substantially, the result of more complete combustion of fuel and the catalytic conversion of carbon monoxide, nitrogen oxides and hydrocarbons into carbon dioxide, nitrogen and water.

Most of the efficiencies achieved in cars are equally applicable to trucks weighing less than 10,000 pounds. This transferability is particularly significant in the U.S., where light trucks now account for about a third of the sales of all passenger vehicles. Light trucks stay on the road longer and consume more fuel than cars do. They thus account for about half of the fuel now consumed in personal transportation.

Fuel consumption can be further reduced at each stage in which chemical energy is converted into mechanical power and then into vehicular motion. Inefficiency begins in the engine, where some energy is lost to friction, waste heat and other factors. In the next stage, involving the transmission and the drivetrain, more of the engine's power is lost to friction. Finally, energy is consumed in overcoming the rolling resistance of the tires and aerodynamic drag.

There are technical means by which to reduce each stage of energy loss. But the solutions vary from region to region, according to the policies of individual governments and the preferences of consumers. The direct-injection diesel, for example, improves on the efficiency of conventional diesels by mixing fuel and air directly in the combustion chamber rather than in a separate prechamber as conventional automotive diesels do. Because, like all diesels, the engine emits substantial quantities of particulates, its future is considered to be dimmer in the U.S. than in Europe. When more rigorous standards go into effect, particulate emissions will be more stringently regulated in the U.S. than in Europe.

Other designs offering better efficiency include the stratified-charge engine, which is receiving renewed interest in Japan. In this engine, precise fuel injection creates a rich mixture of fuel and air near the spark plug so that the spark can cause ignition, and it creates a lean mixture elsewhere in the combustion chamber. Such engines are thought capable of reducing fuel consumption by one fifth, compared with conventional gasoline engines. Yet because their oxygen-rich exhaust prevents catalytic converters from reducing nitrogen oxides to nitrogen, stratified-charge engines still cannot meet the toughest emissions standards.

A particularly interesting application of stratified-charge technology, now attracting worldwide attention, is seen in the modern two-stroke engine (see Figure 5.4). Older versions of this compact design have generally been rejected because they pollute excessively, but careful control of the combustion promises to make them cleaner. If the two-stroke engine can meet projected emissions standards, its low weight would make it very appealing.

Figure 5.1 SNARLED TRAFFIC in Paris shows the results of unbridled growth in the number of cars and trucks. Not only do such increases hasten the exhaustion of limited reserves of oil, they also foul the air and contribute to greenhouse warming. Traffic jams compound the waste and the pollution by causing vehicle engines to idle for extended periods.

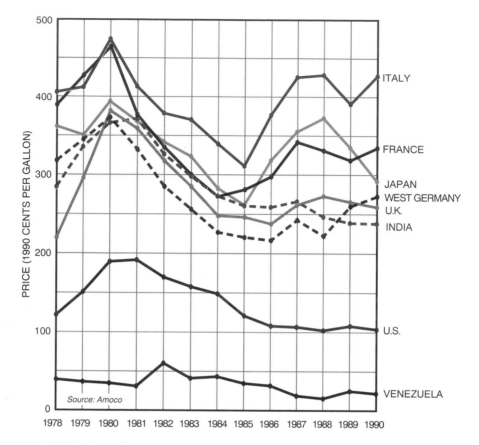

Figure 5.2 RETAIL PRICE of gasoline varies mostly because of differences in government surtaxes. The U.S. imposes the lowest surtax among the five industrialized countries shown here. Of the two developing nations, Venezuela, an oil exporter, prices gasoline near the cost of production; India, an oil importer, taxes gasoline heavily.

The second stage of energy loss occurs in the transmission. Here the challenge is to keep the engine under high load for as much of the time as possible while maintaining the speed the driver has specified. High-load operation—in which most of the engine's power is utilized—is efficient; partial-load operation (during idling, for example) is highly wasteful. One can approach high-load operation by adding more gears or by switching them into their optimal regimes more of the time, with the aid of a computer. Alternatively, one can add the equivalent of an infinite number of gears by transmitting power via a belt drive or other smoothly variable devices. Although such continuously variable transmissions are now available, they can currently be used only in small cars. Several automotive compa-

nies are exploring ways to apply them to large cars as well.

Further energy is lost in overcoming the rolling resistance of the tires and the vehicle's inertial resistance to acceleration. Such losses scale with weight, so that, roughly speaking, a reduction of 200 pounds typically improves fuel economy by nearly 5 percent. Weight can be saved either by a change in the vehicle's design or by substitution of light materials for heavy ones.

Historically, most of the weight-saving changes have involved design (see Figure 5.5). After the first oil crisis in 1973, for example, U.S. automakers reduced weight by shrinking ornamental features such as fins. Perhaps the one fundamental change

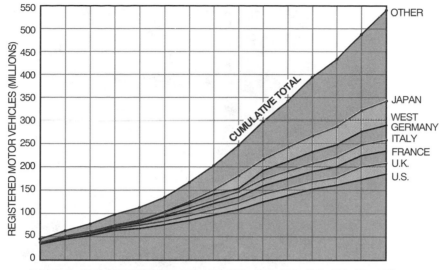

Figure 5.3 GLOBAL FLEET of motor vehicles has grown 12-fold since 1945, increasing oil consumption and air pollution. Most of the growth has been in the industrialized countries, but in the future it will center in Eastern Europe and the developing countries.

the general public has noticed was the widespread transition to front-wheel drive, combined with a transverse-mounted engine. This configuration obviates the need for a "hump" bisecting the passenger compartment, providing more interior space in small and medium-size cars. The design is now being applied to large cars as well, although some problems remain in accommodating large engines and in providing necessary traction. Future design changes to save weight will center on the engine.

Substituting lightweight materials for standard-grade steel can also save more than 100 pounds per car. In the U.S. the substitution of plastics has received more attention than elsewhere, and it probably will continue to do so. Already this modification is seen in several American car and van models. In Europe, on the other hand, concern about the recyclability of plastic has led automakers to put more emphasis on aluminum and high-grade steel.

Aerodynamic drag, another source of friction, increases exponentially with the speed of the vehicle (as does wind noise, an annoyance to passengers). This problem was first perceived in Europe, where speed limits are much higher than in Japan and the U.S. Today, however, manufacturers throughout the world have become interested in aerodynamic

design because it conserves energy in an inexpensive and stylish manner.

The potential for improving fuel efficiency is great, as evidenced by the plethora of test vehicles that were developed, primarily in Europe, during the late 1970's and throughout the 1980's. Because most of the vehicles were intended for research on energy efficiency alone, additional work is needed to incorporate the safety, environmental and performance qualities necessary to make the cars suitable for mass production. Nevertheless, the vehicles do show that lightweight materials, new engine designs, improved transmissions and other factors can increase the fuel economy of individual cars considerably.

A small, two-cylinder diesel engine with an advanced fuel-injection system propels Volkswagen's Eco-Polo, a test car designed for urban commuting. The car incorporates a device, called a glider automatic, that shuts down the engine when the vehicle is decelerating and restarts it automatically when the driver steps on the accelerator. These features yield a combined city/highway fuel economy of 62 mpg, about twice that of most European cars. The Eco-Polo employs an exhaust filter and a special iron-based fuel additive to remove the particulates

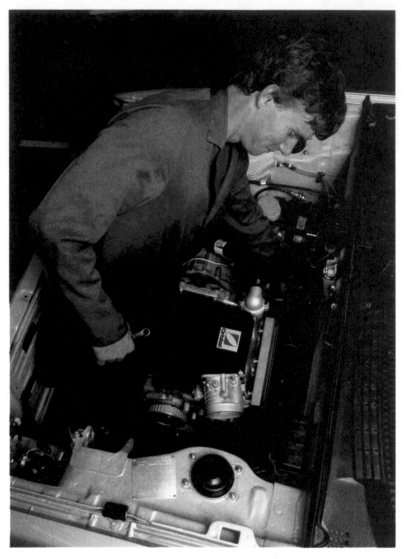

Figure 5.4 TWO-STROKE ENGINE developed by the Orbital Engine Company in Australia is so compact that it leaves enough room in an engine bay to accommodate a man. If the design can be made to meet emissions standards, it will help conserve fuel.

that generally result from the combustion of diesel fuel. Consequently, its particulate emissions are lower than the California standard, the strictest in the world. It must be noted, however, that this additive has not yet been approved as safe by environmental regulators.

Volvo's LCP 2000 is one of the most unusual fuel-efficient research vehicles because it was developed with consumer and production criteria in mind from the start. It was designed to ensure that its passengers survive a head-on crash at 35 miles per hour (mph), a performance that exceeds the U.S. crash standard of 30 mph. Moreover, the car is designed to be assembled from modular components. The ease of assembly would offset much of the added cost of the advanced materials and technologies that enable the car to attain 63 mpg in the city and 81 mpg on the highway.

These test vehicles demonstrate that fuel efficiency can be improved substantially, albeit at greater cost and technical complexity. Even greater

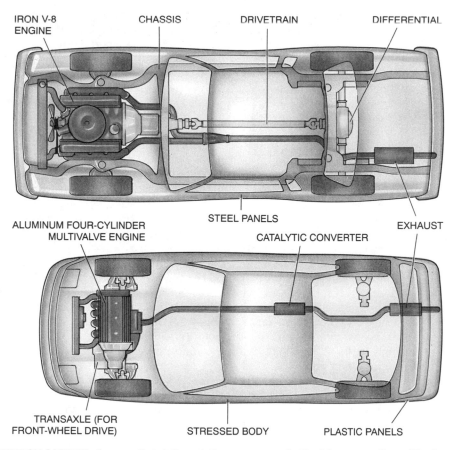

IRON V-8 ENGINE · CHASSIS · DRIVETRAIN · DIFFERENTIAL · STEEL PANELS

ALUMINUM FOUR-CYLINDER MULTIVALVE ENGINE · CATALYTIC CONVERTER · EXHAUST · TRANSAXLE (FOR FRONT-WHEEL DRIVE) · STRESSED BODY · PLASTIC PANELS

Figure 5.5 ENERGY-SAVING changes that followed the oil crises of the 1970's are seen in this comparison of automobile designs. A typical midsize car in the early 1970's (*top*) had a large, iron engine, rear-wheel drive, steel body panels and a chassis for structural rigidity. A comparable car now (*bottom*) has a small, multivalve aluminum engine, compact front-wheel drive, plastic panels and a fully stressed (monocoque) body for rigidity. As a result of such changes, the fuel efficiency of U.S. cars doubled.

efficiency can be achieved but only by sacrificing comfort and vehicle performance. Economic conditions, however, have discouraged manufacturers from bringing their fuel-saving ideas to market, in the form of special, fuel-economy models. The real price of oil has fallen to historic lows in some countries, and so consumers today tend to care less about a car's fuel economy and more about its power and comfort. At the same time, the recent (and, we believe, vanishing) soft market for oil has caused governments to lose interest in the programs they started in the 1970's to encourage the development and purchase of fuel-efficient vehicles. (These programs included the backing of research and the establishment of fuel-economy targets.)

Furthermore, manufacturers will hesitate to develop fuel-efficient vehicles that cost more than customers are prepared to pay or that contain technologies with which consumers are uncomfortable. Volkswagen, for example, would not consider mass-producing a car that incorporates a glider automatic without first testing driver acceptance of the engine's intermittent operation. Consumers might well reject this feature, since it involves a certain reduction in the perception (if not the reality) of safety.

Improvements in fuel economy alone cannot solve the problems of oil availability and clean air. Equally important in the long run is the use of alternative fuels, preferably those whose production and combustion add no net carbon dioxide to the atmosphere. Only three fuels meet this ideal criterion: hydrogen, if produced from nonfossil-fuel sources; biomass, which consumes as much carbon dioxide during photosynthetic growth as it releases during combustion; and electricity, if it is generated from nonfossil-fuel sources.

All three are still impractical. Hydrogen gas has a low density of energy, which limits the range and payload of a hydrogen-powered vehicle. It also suffers from a low density of power, which limits its performance (ability to accelerate). Hydrogen is also inconvenient to store and distribute. Daimler-Benz and BMW are working on hydrogen-powered cars but are not expected to bring them to market until well into the next century.

Electric vehicles have low energy and power densities, too, plus the disadvantage of a long recharging period. Yet they operate quietly, waste no fuel when standing still and can even recover some of the energy normally lost in braking by using their motors as generators (a process called regenerative braking). Despite the problems with electric vehicles, they are increasingly seen as attractive for urban conditions because they emit no pollutants at all. Several companies have already begun to market some electric cars and vans, but they are expected to remain confined to niche markets where long-distance driving and quick refueling are not essential.

Ethanol (grain alcohol) is the major biomass fuel today, but in current market conditions it remains significantly more expensive than gasoline. In the U.S. ethanol is produced from corn for use as a gasoline additive. In Brazil manufacturers produce it from sugarcane and blend it with water to make hydrous ethanol. At the height of the program, 90 percent of all new cars sold in Brazil were powered by ethanol, but the ethanol production has been drastically curtailed in recent years, largely because of financial difficulties. If more efficient biomass-to-alcohol conversion processes prove practical, ethanol and methanol (wood alcohol) might become economic. In addition, European countries are investigating rapeseed oil as an alternative fuel.

Unfortunately, the widespread use of biomass fuels is constrained by the size of the resource base from which they are produced. Europe and Japan simply do not have enough land on which to grow large amounts of biomass for automotive fuel. Even in the U.S. some analysts have estimated that biomass available today—crop stubble, forage crops, wood chips, garbage and peat—could provide no more than 20 to 30 percent of the energy currently required for transportation. Plantations whose whole yield was dedicated to biomass could provide the balance but possibly at the expense of farms and forests (which absorb carbon dioxide).

Other analysts propose the adoption of transitional technologies pending the perfection of hydrogen, electricity or biomass technologies. In the U.S. considerable attention has been paid to methanol, which can be produced from natural gas or coal (see "The Case for Methanol," by Charles L. Gray, Jr., and Jeffrey A. Alson; SCIENTIFIC AMERICAN, November, 1989). Methanol produces smaller quantities of noxious emissions than gasoline does, and it can be produced from indigenous resources, thereby reducing U.S. dependence on imported oil.

Critics argue that methanol produced from natural gas releases as much carbon dioxide as gasoline does and that domestic reserves of natural gas would quickly be exhausted. After the reserves disappeared, the U.S. would be as dependent on politically unreliable fuel suppliers as it is today. Coal, they add, can be converted to methanol only at the cost of even greater emissions of carbon dioxide. Advocates counter that methanol-fueled engines have a greater potential for improvements than gasoline engines do (for example, they can function at higher levels of compression than have heretofore been attempted). They also contend that biomass, rather than natural gas, can eventually serve as a feedstock for methanol.

The energy contained in natural gas might be exploited more efficiently if the fuel were burned directly instead of being converted to methanol. Indeed, in New Zealand, Italy and other countries, many vehicles have been converted to natural gas. The fuel—which largely consists of methane—produces about 20 percent less greenhouse gas than gasoline does (in terms of equivalent quantities of carbon dioxide). But natural gas has a low energy density and must be stored under compression in heavy, bulky tanks, limiting the range and payload of vehicles. Moreover, existing distribution systems are designed for liquids and would require fundamental changes to accommodate natural gas.

Another transitional concept would combine gasoline and electricity to exploit their respective advantages of power and cleanliness. Volkswagen has designed such a hybrid. The test vehicle has a diesel engine, a small electric motor, a sodium-sulfur battery (which operates at 300 degrees Celsius) and a clutch that yokes the motor to the engine. When the accelerator pedal is pressed less than a third of the way down — as it would be during most urban travel — the electric motor powers the car. Further pressure on the pedal engages the clutch, so that the electric motor — serving now as a flywheel —starts the engine. The electric motor can also serve as a generator to recharge the battery. When driven according to the mix of city and highway travel typical of Europe. Volkswagen's hybrid travels almost 100 miles on a gallon of diesel and 25 kilowatt-hours of electricity. If the electric charge comes from sources other than fossil fuels, 60 percent of the carbon dioxide emissions normally emitted by a vehicle of this size would be eliminated.

Many barriers stand in the way of the speedy adoption of alternative fuels and technologies. The greatest of these obstacles is uncertainty. No one yet knows what the next technology will be. In such an environment, few manufacturers are willing to commit themselves to new fuels or to developing vehicles that would use them. Another obstacle is presented by alternative fuels' cost, which is much higher than the current cost of oil. Nevertheless, the necessity of controlling smog in some cities may well bring about large-scale experimentation with new vehicles and energy sources.

So far we have concentrated on vehicles and fuels; there are other ways to save energy and reduce pollution. The transportation system can itself be reformed. Roads, parking and regional traffic patterns can be redesigned to ease congestion and make much routine commuting unnecessary. Congestion is not merely annoying; it also wastes fuel and as a result increases air pollution.

Perhaps the best-known attempt to improve traffic flow and safety is the PROMETHEUS project, which is funded by European automakers. A version is currently being tested in Berlin, where 200 traffic lights bear infrared transmitters whose signals can be received by an experimental fleet of 900 cars. At the start of a trip, the driver enters the desired destination into a computer. Each time the car passes a transmitter, the computer receives information about traffic conditions, which it then processes to calculate the quickest route. Test results are expected by the end of this year.

More advanced systems may one day be adopted. Electronics might augment human reaction speeds, making it possible for drivers to maintain shorter separations between their cars without compromis-

ing safety. More cars could fit on the road without interrupting the steady flow of traffic. This facilitation of movement could save up to 20 percent of the fuel consumed and could double or triple the carrying capacity of the roads.

Congestion can also be eased by limiting inner-city parking, so that drivers park in the outskirts and take shuttle buses to work (see Figure 5.6). The system already operates on a voluntary basis in many U.S. cities. Commuting distances could be shortened, too. As urban populations grow, people are often compelled to move farther away from their jobs, often relocating beyond the reach of mass-transit systems (if they exist at all). If industrial and commercial enterprises were to relocate nearer the people they employ, commuting distances would be greatly reduced. The Netherlands, seeking to capture such potential savings, recently developed national zoning laws to prevent commercial and residential construction in places that are not near major public transit terminals.

Finally, the substitution of mass transit for some car travel could slow energy consumption per capita. Some countries, however, cannot exploit mass transit as easily as others can. In the U.S., for example, mass transit accounts for only 6 percent of all passenger travel, and the expansion of the system would be very costly in relation to the amount of energy saved. Even if the U.S. were to triple the size of its mass-transit system, so that it carried the same fraction of passengers as Europe's systems do, and even if the enlarged system were filled to capacity, energy use by private vehicles would decline by only 10 percent.

But in West Germany, where 15 percent of all passenger travel is by mass transit, and Japan, where that figure is 47 percent, the mass-transit infrastructure is well developed, and intercity distances are relatively small. Because both countries stand to benefit even more from the improvement of the system, they are investing heavily in the development of trains that can travel more than 200 mph.

An expanded mass-transit system can work only if commuters would rather use it than drive. Present conditions, however, tend to reinforce the preference for private cars. Mass transit is sometimes inaccessible, uncomfortable and dangerous. In addition, buses and trains run infrequently, particularly during off-peak periods. Some efforts are being made to make mass transit more competitive.

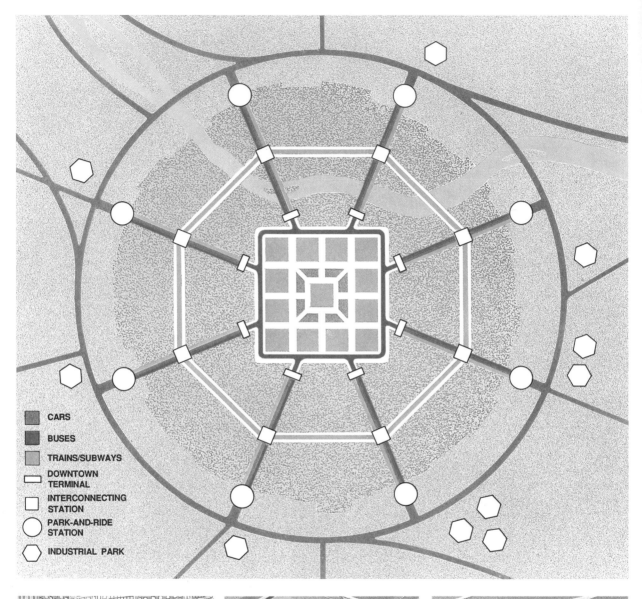

CARS

BUSES

TRAINS/SUBWAYS

DOWNTOWN
TERMINAL

INTERCONNECTING
STATION

PARK-AND-RIDE
STATION

INDUSTRIAL PARK

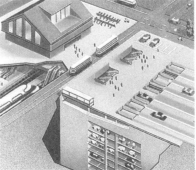

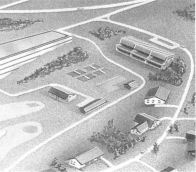

Washington, D.C., for example, has built commuter parking lots at the outlying terminals of its subway system.

Carpooling is one energy-saving option that offers much of the efficiency of mass transit and some of the flexibility of private vehicles. This option is particularly appealing in the U.S., where cars carry an average of 1.7 persons. If all passengers there traveled in carpools of four persons each, gasoline consumption would fall by about 45 percent. When gasoline prices dip, however, additional incentives are required to promote carpooling. Washington, D.C., for example, has reserved express highway lanes for vehicles carrying at least three people. The resulting demand for carpools is so strong that suburban drivers often form lines at convenient points in the suburbs, where they wait to pick up enough passengers to qualify for the lanes.

Walking and bicycling are other attractive alternatives to driving. They save energy and reduce air pollution and add no net carbon dioxide to the environment. But human-powered travel faces stiff consumer resistance wherever safety and comfort are at stake. In most urban regions in the OECD countries, cyclists are either mixed with motorized traffic or assigned to lanes that lack protective barriers. In suburban areas, pedestrians must often do without sidewalks and even crosswalks.

The Netherlands, one of the most densely populated OECD countries, has worked hard to create incentives for the use of bicycles. The government has set aside paths and parking spaces for bicycles, established rent-a-bike facilities at railroad stations and allowed train passengers to bring their bicycles on board. As a result, bicycles carry fully 9 percent of the country's commuters. In some cities, they account for more than 40 percent of all passenger trips.

Although a large majority of the vehicles on the road today are in the OECD countries, most of the increase in the world's fleet over the next 50 years is likely to occur in Eastern Europe and the developing countries. There are many reasons to keep the concomitant increase in oil consumption as low as possible. Countries that lack oil will have to spend scarce foreign exchange to import it; countries that now export oil will have to divert some of it to their expanding domestic markets. In either case, the supply of investment capital for development will dwindle, and the external debt could well grow. At the same time, regional air quality will undoubtedly deteriorate, as it already has in many traffic-congested cities in the developing world.

Developing nations do have one crucial advantage over developed ones: they can head off many problems of unplanned industrialization before they become intractable. Because developing countries have not yet institutionalized the private car to the degree seen in the OECD nations, they are still in a position to create mass-transit systems that people will want to use.

Curitiba, capital of the state of Paraná, in southeastern Brazil is a famous model of how planning can avert the disadvantages of wasteful fuel consumption and gridlock. The city's transportation system revolves around five radial express lines reserved exclusively for buses. These arteries are connected by interdistrict lines, and the whole system is linked to neighborhoods by feeder lines. Land-use ordinances of residences and businesses near bus stops. As a result, Curitiba enjoys one of the highest rates of motor vehicle ownership per capita and one of the lowest rates of fuel consumption per vehicle in Brazil. A comparatively large number of people have cars, but most of them prefer mass transit for routine urban travel.

Of course, even with the aggressive pursuit of mass transit, the demand for cars in Eastern Europe and the developing countries is likely to increase — but not as fast as it did in regions that industrialized earlier. It is therefore crucial that the cars they import be as efficient as possible. Furthermore, as developing nations establish domestic automotive industries of their own, it is of critical importance that they make efficient products.

Because of the increasing consumption of energy by the transportation systems of Eastern Europe and the developing countries, air quality and the balance of payments are deteriorating hand in hand. Many of these countries might consider developing indigenous alternative fuels, just as Brazil did in the

Figure 5.6 IDEAL URBAN DESIGN links a city's center to peripheral areas via express highways and railroads (*top*). Downtown terminals are served largely by buses or trains, although some commuters drive to the city center, where parking is available (*bottom left*). Suburbs connect to mass-transit systems at park-and-ride stations, which have automated underground garages (*middle*). Some commuting is eliminated by locating businesses near employees' homes (*bottom right*).

late 1970's, when skyrocketing oil prices and plummeting sugar prices devoured that country's supply of foreign exchange.

Even though most developing countries are not as well endowed with biomass feedstock as Brazil and may not be able to produce enough fuel to run all their vehicles, biomass can still replace a significant fraction of these imports. Moreover, quite a few of these countries possess unexploited reserves of natural gas. Thailand, Indonesia and Argentina, for example, are already testing cars that run on domestic natural gas. Before any alternative-fuel programs are launched, however, all the risks and benefits must be assessed comprehensively.

But the effort to rethink transportation systems must begin in the OECD countries, which created the problems of waste and pollution. This effort must first concentrate on light vehicles, which predominate in overall transportation energy use. Progress in addressing this sector can be made only with the cooperative leadership of national governments. The OECD members can perhaps best begin their task by signing a protocol on the lowering of carbon dioxide emissions from road vehicles. That step could serve as the model for other countries whose transportation sectors are only now beginning to expand.

Energy for the Developing World

By mixing efficient end-use technologies with modest increases in generating capacity, developing countries can affordably obtain the energy they need without ruining the environment.

· · ·

Amulya K. N. Reddy and José Goldemberg

If current trends persist, in about 20 years the developing countries will consume as much energy as the industrialized countries do now. Yet their standard of living will lag even farther behind than it does today. This failure of development is not the result of a simple lack of energy, as is widely supposed (see Figure 6.1).

Rather, the problem is that the energy is neither efficiently nor equitably consumed. If today's most energy-efficient technologies were adopted in developing countries, then only about one kilowatt per capita used continuously—roughly 10 percent more than is consumed now—would be sufficient to raise the average standard of living to the level enjoyed by Western Europe in the 1970's.

This fact seems surprising because the developing world is one of stark contrasts. Although developing countries vary tremendously, they are all "dual societies," consisting of small islands of affluence in vast oceans of poverty. The elite minorities and the poor masses differ so much in their incomes, needs, aspirations and ways of life that, for all practical purposes, they live in two separate worlds. Consequently, the elites and the poor differ fundamentally in their use of energy. The elites emulate the ways of life prevalent in industrialized countries and have similar patterns of luxury-oriented energy

use (see Figure 6.2). In contrast, poor people are preoccupied with finding enough energy for cooking, obtaining water and other activities essential to survival.

Much of the energy for agriculture, transportation and domestic activities in developing countries comes from human beings and draft animals. Energy also comes from other sources, particularly biomass in the form of fuelwood, animal wastes and agricultural residues. Fuelwood, in fact, is the dominant source of energy in rural areas, and cooking is the most energy-intensive activity. These biological sources of energy are often described as noncommercial because they are not purchased: for example, in rural areas, the women and children usually gather twigs and branches for cooking fuel instead of buying wood.

Because most of the population in a developing country is poor and depends largely on noncommercial sources of energy, per capita use of commercial energy is much lower than in an industrialized country. Even when noncommercial sources are considered, developing countries' level of total energy services—heating, cooling, lighting, mechanical power and so on—is much lower, particularly because the end-use devices for delivering those services are so inefficient.

The great disparities between the elites and the masses, and between the industrialized and developing world, have led to widespread pressure for stepping up the level of energy services available to those who lack them. To date, decision-makers have interpreted the pressure as a mandate for escalating commercial energy consumption. The implementation of this interpretation, coupled with population growth, had led to an almost linear rise in energy consumption in the developing world for two decades. As we shall discuss, such increases are unsustainable, and a new view of the energy problem is essential.

In the capital-constrained modern world, it is becoming more difficult for developing countries to procure the capital they need to expand their production of energy. The World Bank quantified the problem at the 1989 World Energy Conference in Montreal by revealing that the capital requirements of developing countries will add up to a trillion dollars during the next decade for the electricity sector alone. Yet the World Bank and other multilateral funding agencies will be able to provide only about $20 billion a year. Within developing nations themselves, the capital demanded by the electricity sector is four to five times what is available.

Aside from its economic implications, the current upward trend in energy consumption has serious environmental implications. At the local level, major consequences of rising energy consumption include the flooding and submergence of forests when hydroelectric dams are constructed, atmospheric pollution and acid rain caused by coal-based thermal-power plants, and deforestation brought about by the high urban (and sometimes rural) demand for fuelwood for cooking.

Higher energy consumption also affects the global atmosphere. Fossil-fuel consumption in developing countries, which today contributes 19 percent of all fossil fuel-derived emissions, is likely to double within 20 years. Deforestation contributes to the increase of carbon dioxide in the atmosphere

Figure 6.1 CHALLENGE of providing more energy services—such as lighting and refrigeration—in developing countries can be met while reducing the environmental risks posed by current consumption patterns. Technologies that deliver the services efficiently are the key. In some cases, electricity can replace fuelwood as a more suitable carrier of energy.

because trees absorb that gas during photosynthesis.

Most often, deforestation occurs to obtain clear land for ranching and agriculture or to obtain wood for lumber and feedstock for the paper and rayon industries. Energy-related causes of deforestation are the submergence of forests by dammed rivers and the collection of fuelwood (logs and charcoal) for industries and urban households. Rural households also sometimes contribute to deforestation when the rate of their wood gathering exceeds the rate at which the forests regenerate.

To appreciate the effect of fuelwood consumption on deforestation, consider the case of Africa. The total amount of fuelwood, including charcoal, consumed there is approximately 300 million tons. If this fuelwood is obtained by felling trees, at least two million hectares (about five million acres) must be deforested every year, which leads to soil erosion, the loss of species and local climate changes. Burning the wood contributes about 4 percent to the worldwide total of carbon dioxide emissions.

Deforestation in developing countries is now responsible for about 23 percent of global carbon dioxide emissions and shows no signs of diminishing (see Figure 6.3). When both fossil-fuel emissions and deforestation are considered, it is clear that even if industrialized countries stabilize their emissions of greenhouse gases, developing nations will degrade the global atmosphere on their own.

Current energy trends cause conflicts within developing countries. Aid agencies, governments, industrialists and other promoters of development push for drastic rises in energy consumption, even though such rises require impossibly large amounts of capital and harm the local environment. Local environment groups then argue that the development process is unsustainable and should be halted. Meanwhile people who are displaced from their homelands because of hydroelectric dams and other energy projects, see themselves as the victims of development rather than as its beneficiaries (see Figure 6.4).

The conflict between the need for development and the need to sustain the planet dominates international discussions of energy issues. To control the degradation of the atmosphere, industrialized countries often seek to limit increases in energy con-

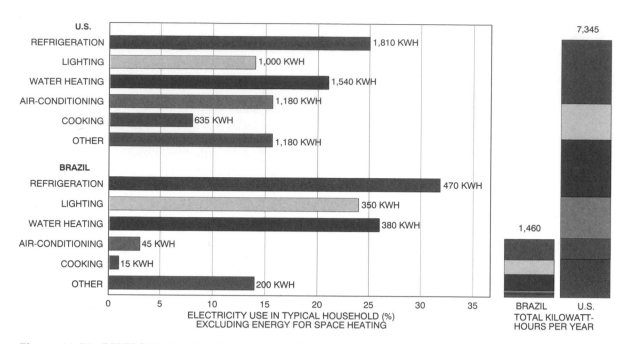

Figure 6.2 IN DEVELOPING COUNTRIES relatively wealthy people tend to emulate the ways of life and energy-use habits of the industrialized world. Electrically wired households in Brazil and the U.S., for example, have generally similar patterns of electricity consumption, even though their levels of total energy consumption contrast more sharply in absolute terms.

sumption by developing countries. Developing countries respond by emphasizing their need to raise their low levels of per capita energy consumption. As long as the two factions are trapped in the conventional paradigm that views energy consumption as an indicator of development, the conflict cannot be resolved without sacrifice.

A new paradigm for energy use is therefore essential. Energy must be viewed not as an end in itself or as a commodity but as a means of providing services. For it is the services, and not the energy, that directly satisfy people's needs: the quality of life in a village depends more on the amount of illumination (measured in lumens), for example, than on the kilowatt-hours of electricity or liters of kerosene consumed for lighting. The extent to which energy services are accessible is therefore the true indicator of the level of development.

Development consequently requires major increases in the per capita level of energy services. Such services, in turn, rely on end-use devices, such as stoves, lighting fixtures and motors, that convert the energy for use. More efficient devices can conserve energy by delivering the same services with less consumed energy or greater services with the same energy (see Chapter 2, "Efficient Use of Electricity," by Arnold P. Fickett, Clark W. Gellings and Amory B. Lovins).

Energy conservation in developing countries must not be achieved through a decrease in services. It must be based on expanding energy services while stabilizing or curtailing energy consumption —for instance; producing more light with the same or fewer kilowatt-hours of electricity. Conservation alone may be adequate for meeting the present needs of developing countries. Development, however, requires industrialization, a process that depends on growth in the production of goods and the performance of commercial services, as measured by the gross domestic product (GDP). Annual GDP growth rates of between 5 and 10 percent have become the standard goals of developing countries, although only a few have achieved such targets.

When Western Europe and North America were undergoing industrialization, their energy con-

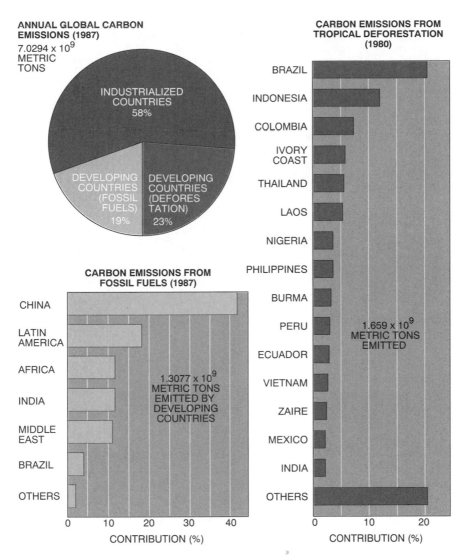

Figure 6.3 FOSSIL-FUEL BURNING and deforestation in developing countries contribute to the buildup of atmospheric carbon in forms such as carbon dioxide, carbon monoxide and soot. Development plans that would expand the share of carbon emissions from these nations pose a threat to the atmosphere and possibly to global climate.

sumption had to grow faster than their GDP's to build infrastructures: roads, bridges, houses and heavy industry. Because of the revolution in materials science that has taken place during the past half-century, however, materials can now be produced with less energy, and smaller quantities of modern materials can replace larger amounts of older ones (see Figure 6.5). Consequently, developing nations can achieve comparable levels of indus-trialization with a lower ratio of consumed energy to GDP growth.

Despite the promise of conservation technologies and advanced materials science, significant increases in energy consumption will probably be vital for development. The developing world needs a balanced mix of end-use efficiency improvements and centralized and decentralized technologies. (Centralized technologies, such as nuclear power,

Figure 6.4 HYDROELECTRIC DAMS can flood prime forests and displace people from their settlements. These problems must be considered in advance by energy planners.

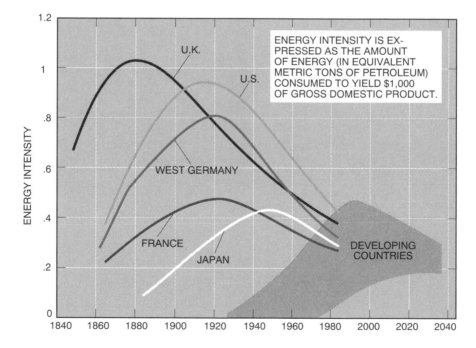

ENERGY INTENSITY IS EXPRESSED AS THE AMOUNT OF ENERGY (IN EQUIVALENT METRIC TONS OF PETROLEUM) CONSUMED TO YIELD $1,000 OF GROSS DOMESTIC PRODUCT.

Figure 6.5 IN INDUSTRIALIZED COUNTRIES the energy intensity (ratio of energy consumption to gross domestic product) rose, then fell. Because of improvements in materials science and energy efficiency, the maxima reached by countries during industrialization have progressively decreased over time. Developing nations can avoid repeating the history of the industrialized world by using energy efficiently.

generate electricity at one location and distribute it widely. Decentralized technologies, such as small hydroelectric dams, produce smaller amounts of energy that are used nearby.)

One quantitative way to identify the components of such a mix is to plot a least-cost supply curve for available technologies. Such a curve makes it possible to weigh the costs and potential energy contributions of each technology and to determine the least expensive combination of technologies for achieving an energy-supply goal.

It is invariably cheaper to save a kilowatt than to generate a kilowatt. Also, generating energy as close as possible to where it will be consumed minimizes transmission and distribution costs. Consequently, many conservation and decentralized technologies find a place in a least-cost mix. Because energy-efficiency improvements are environmentally benign, the resulting technology mixes can advance development without jeopardizing sustainability.

Because improvements in efficiency make it possible for the GDP to rise while energy consumption remains the same, technology mixes that include energy-efficiency measures reduce the coupling of energy and the GDP. As a result, the required annual investment in energy diminishes and becomes more affordable.

The validity of weakening the coupling of energy and the GDP has been confirmed by studies of the long-term evolution of the energy-GDP ratio for many countries. The studies show that the ratio decreased steadily except when the countries were establishing their heavy-industry infrastructure. Several factors caused the decline: a saturation of the demand for consumer goods in industrialized countries, a shift of economic activity away from heavy materials-processing industries toward services and the revolution in materials science.

The revised paradigm highlights the importance of scenarios that focus on development objectives and that find opportunities to improve the efficiency of end uses for energy. In other words, planners should construct what we call development-focused, end use-oriented, service-directed (DEFENDUS) scenarios that incorporate conservation and renewable sources into a least-cost mix.

One of (Reddy) has helped to construct a DEFENDUS scenario for the electricity sector of the state of Karnataka in southern India. The scenario was a response to recent efforts at electricity planning—in particular, the committee report "Long Range Plan for Power Projects [LRPPP] in Karnataka," dated May, 1987—which were clear-cut examples of the failure of conventional consumption-obsessed, supply-biased approaches to energy planning (see Figure 6.6).

To meet its projected goal of producing 47,520 gigawatt-hours of energy and almost 9.4 gigawatts of power by the year 2000, the LRPPP demanded that Karnataka should spend the astronomical sum of about $17.4 billion—an amount roughly equal to 25 times its annual budget. The state would need to build an extensive energy infrastructure, construct massive centralized power-generating facilities (including a one-gigawatt coal-based thermal-power station and about two gigawatts or nuclear power), raise funds from the World Bank and the central government, divert at least 25 percent of its budget to power, and appeal to private industry to build up generating capacity. Despite these measures, the LRPPP projected that energy shortages would continue into the next century. Such a conventional energy plan was not a solution; it was an exercise in profligacy.

The DEFENDUS scenario for Karnataka turns out to be as promising as the LRPPP is gloomy. It calls for the installation of electric lights in all homes in Karnataka, the use of electric irrigation pumps up to a limit imposed by the groundwater potential, the establishment of decentralized energy centers in villages and the promotion of industries to increase employment. The DEFENDUS scenario nonetheless estimates that its requirements for energy and power in 2000 will be only about 38 and 42 percent, respectively, of the LRPPP demands.

Fifty-nine percent of the reduction in the energy requirement is achieved through a better-directed development focus: a society that improves the lot of the poor prudently needs less energy than one that makes no such dent in poverty. The other 41 percent stems from simple efficiency improvements and the substitution of one energy carrier for another. These measures include replacing inefficient motors and incandescent bulbs with more efficient motors and compact fluorescent lamps, substituting solar-powered water heaters and liquefied-petroleum gas (LPG) stoves for electric water heaters and electric stoves, and retrofitting irrigation-pump systems with frictionless foot valves and better piping.

Along with these efficiency improvements, the DEFENDUS scenario also calls for a mix of technologies that substitute for electricity (such as solar-powered water heaters), decentralized power sta-

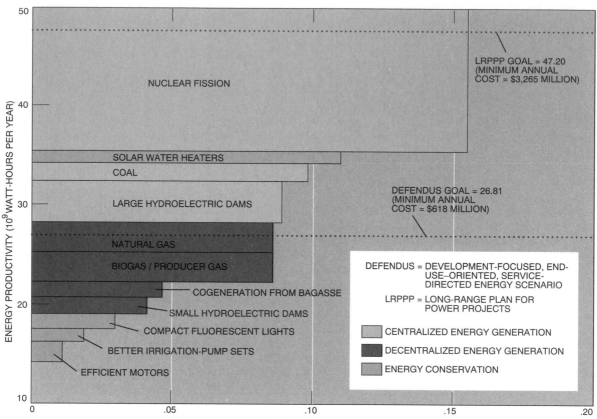

Figure 6.6 TWO ENERGY PLANS proposed for the state of Karnataka in India differ greatly in their energy and capital requirements. Although the DEFENDUS plan has a much lower energy requirement than the LRPPP, it calls for providing greater services. The costs and energy contributions of various technologies for conserving and generating energy are plotted to calculate the least expensive means of reaching the DEFENDUS and LRPPP energy goals for the year 2000. (The costs of setting up transmission and distribution lines are included in the figures for the centralized technologies.) The DEFENDUS goal could be met without introducing controversial technologies such as nuclear power and large hydroelectric dams.

tions (such as small hydroelectric plants and biomass-based rural energy centers) and conventional centralized generating plants. Because the energy requirement has been drastically reduced, the most environmentally controversial centralized technologies—nuclear-power plants, coal-based thermal-power plants and large hydroelectric dams—can generally be avoided.

The estimated cost of the DEFENDUS scenario would be only one third that of the LRPPP, which means that the DEFENDUS plan would result in a much lower per unit cost for energy. Even if 10 years were needed to introduce the efficiency im-

provements, the DEFENDUS plan would be able to provide more energy sooner than the LRPPP. And the DEFENDUS scenario would release only about .5 percent as much carbon dioxide into the atmosphere every year. Karnataka has rejected the LRPPP; the DEFENDUS scenario is only now approaching consideration.

The DEFENDUS scenario for Karnataka is not intended as a universal recipe for an inexpensive, fast, environmentally benign energy plan. Because energy-consumption patterns and resource endowments in other regions differ from those in Karnataka, the conservation potentials and the least-cost supply mixes are bound to vary. Nevertheless, the quantitative approach that underlies the DEFENDUS scenario is applicable in other developing countries, and therein lies its strength.

As promising as the DEFENDUS scenario for Karnataka is, it is technologically quite timid and conservative, entirely based as it is on proved, off-the-shelf technologies. Moreover, the plan does not take into account the additional advantages of using the most energy-efficient technologies to build a better infrastructure.

The large amounts of capital that industrialized countries have already invested in their infrastructures discourage further investments in energy-efficiency upgrades. In developing countries, however, many of the vital industries, buildings, roads and transportation systems are not yet in place. Developing nations therefore have the opportunity to adopt more energy-efficient technologies even before they have been adopted widely in industrialized countries. Such "technological leapfrogging" must become an integral component of DEFENDUS energy strategies.

The alcohol program in Brazil—an innovative response to the oil crisis of the 1970's—is an excellent example of technological leapfrogging. Faced with a growing deficit in its trade balance caused by an enormous jump in petroleum prices, Brazil decided to substitute pure ethanol and gasohol (mixtures of ethanol and gasoline) for gasoline in automobile engines. Aside from the economic considerations, ethanol has a higher octane ratio and other technical advantages over gasoline.

The production of ethanol from fermenting sugarcane juice rose from 900 million liters in 1973 to 4.08 billion liters in 1981, of which 1.88 billion liters were turned into hydrated ethanol (91 to 93 percent ethanol plus water); the remaining 2.2 billion liters became anhydrous ethanol mixed with 20 percent gasoline (see Figure 6.7). In 1989, 12 billion liters of ethanol replaced almost 200,000 barrels of gasoline a day in approximately five million Brazilian automobiles. The alcohol industry created 700,000 jobs. The excellent performance of ethanol-fueled automobiles significantly improved the quality of the air in polluted metropolises such as São Paulo and Rio de Janeiro.

Above all, Brazil, a developing country, established an entire fuel cycle—from an energy source (sugarcane) to end-use devices (alcohol-fueled automobiles)—that does not exist in industrialized countries. Like children vaulting over their playmates, Brazil has leapfrogged technologically over the industrialized countries.

The average cost of ethanol produced in the southern region of Brazil is currently 18.5 cents per liter. At this price, ethanol could compete successfully with imported oil if the international price of oil was $24 per barrel. When the oil price fell in the mid-1980's, however, the Brazilian ethanol program faced a serious economic crisis and had to be subsidized by the government.

The country's general economic condition provoked domestic pressure to reduce the cost of ethanol further while also removing the subsidies. This pressure led to major efforts to improve the productivity and economics of sugarcane agriculture and ethanol production. As a result, ethanol costs have fallen 4 percent a year. Brazilian ethanol distilleries have become the best in the world and compete strongly in the international market.

The effective cost of ethanol can be decreased even more if bagasse, the residue left after sugarcane is crushed and drained of its juice, is burned efficiently to make steam for generating electricity. Today low-pressure steam-turbine systems for generating electricity can produce about 20 kilowatt-hours per ton of sugarcane. High-pressure steam turbines could generate up to three times as much electricity, and gas turbines could yield 10 times as much. Such improved cogeneration facilities could make complexes that distill ethanol from sugarcane even more enticing as energy-exporting enterprises.

In short, if DEFENDUS scenarios are not limited to currently available technologies but also resort to technological leapfrogging, they are likely to become even more attractive.

Four national priorities that constitute the essence of sustainable development emerge from the DEFENDUS approach. First, developing countries should assign overriding importance to satisfying the basic energy needs of their populations. Second, they must overcome the severe capital constraints frustrating conventional energy futures. Third, resources must be consumed more efficiently. And fourth, energy production and use must be managed to minimize local and short-term environmental impacts.

Should these priorities be ranked in the above order? The answer would be no if one were influenced by the example of the industrialized countries, where environmental concerns intensified be-

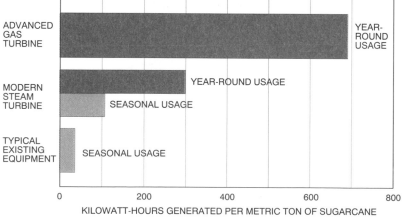

Figure 6.7 ETHANOL DISTILLER-IES in Brazil (*top*), the most efficient in the world, convert the juice from sugarcane into alcohol that local automobiles burn for fuel. Better equipment for generating electricity from steam produced by burning sugarcane residues (bagasse) could turn the sugar factories into energy exporters (*bottom*).

ADVANCED GAS TURBINE — YEAR-ROUND USAGE

MODERN STEAM TURBINE — YEAR-ROUND USAGE / SEASONAL USAGE

TYPICAL EXISTING EQUIPMENT — SEASONAL USAGE

0   200   400   600   800
KILOWATT-HOURS GENERATED PER METRIC TON OF SUGARCANE

fore the problems of local poverty were solved. On the other hand, the answer would be yes if one notes that the poor stand to benefit the most from environmental improvements: the poor are in worse health and cannot flee from damaged environments as easily as the rich can. To the poor, however, survival is a problem of such urgency that, if necessary, they will survive at the expense of the environment. As the late Indian Prime Minister Indira Gandhi proclaimed in Stockholm at a conference on the environment, "Poverty is the greatest polluter!"

Clearly, both the improvement of the environment and the reduction—if not the eradication—of poverty are important and deserve simultaneous attention insofar as it is possible. Such a thrust emerges naturally from a view of development as an environmentally sound, self-reliant process that is directed to meet certain needs.

If putting national priorities ahead of an international obligation to preserve the atmosphere seems wrong, it is only because conventional development plans based on energy consumption compel a trade-off between these priorities. The worst trade-off is one in which developing countries are urged to protect the global ecosystem at the expense of urgent national development tasks. The only likely response to such recommendations is that industrialized countries must assume responsibility for warding off a climatic catastrophe because they are primarily guilty of building up the greenhouse gases and because it is in their own self-interest.

Fortunately, such a head-on collision between industrialized and developing countries is unnecessary. The only strategies for averting disastrous climate changes likely to find acceptance in developing countries are those that simultaneously address development priorities. In an energy-efficient future, development plans can meet people's needs while incidentally helping to alleviate the world's environmental woes.

Energy futures compatible with a sustainable world and sustainable development are within our grasp. The choices that we propose require imaginative political leadership, but they represent far less difficult and hazardous options than those demanded by conventional approaches.

Implementing changes in the energy system involves actors at every level, beginning with individual consumers. Consumers who do not obtain their energy efficiently fall into three categories: the ignorant, the poor and the indifferent. The first consists of people who do not know, for example, that cooking with LPG is more efficient than cooking with kerosene. They can be educated to become more energy efficient.

The second category consists of those who do not have the capital to buy more efficient appliances, which usually have higher initial costs. An Indian maid may know that her employers spend less money cooking with LPG than she does with kerosene, yet she may not be able to switch to a gas cooker because LPG stoves are about 20 times more expensive than kerosene stoves.

To help consumers who are deterred by high initial costs, utilities or other agencies should help finance the purchase of efficient equipment with a loan that can be recovered through monthly energy payments. Alternatively, a utility can lease energy-efficient equipment (see Figure 6.8). A consumer's savings in energy expenditures can exceed the expenses of loan repayments and new energy bills. In principle, this method of converting initial costs into operating expenses can be extended to commercial and industrial customers as well, thereby improving efficiency and modernizing equipment at the same time.

The third category of consumers consists of those with little incentive to raise their energy efficiency because their energy costs are so small or because the costs are almost unaffected by efficiency changes. For example, in the U.S. there is a range of automobile fuel efficiencies—"the plateau of indifference"—within which the total operating costs for the vehicles are nearly the same.

Enticing those consumers to make energy-efficiency improvements will depend on intervention at higher levels. Governments may have to legislate higher efficiency standards or initiate regulations for the efficiency of end-use devices, as the U.S. Government did when it set fuel-efficiency targets for its automobile fleet.

To improve the energy systems in developing countries, the traditional attitudes of utilities, financial institutions and governments will have to change. They must establish methods for converting the initial cost of efficient systems into an operating expense.

Huge investments were necessary to build the existing energy systems throughout the world. They were financed with guaranteed state loans, public loans and direct government investments—all of which permitted the investments to be repaid over long periods and at much lower interest rates than

Figure 6.8 UTILITIES can encourage consumers to use more efficient appliances, such as gas stoves, by leasing the equipment or offering cheap loans for their purchase.

were prevalent in the business world. In developing countries, where capital is scarce, most investments are made by the government and are usually backed by loans from international banking institutions such as the World Bank.

Federal regulations in most nations guarantee that many utilities receive a net return of about 10 percent a year on capital invested in generating more energy. Investments in energy production therefore run virtually no risk at all. In contrast, investments in efficiency measures venture and must be repaid to commercial banks over relatively short periods at high interest rates.

Yet efficiency improvements invariably involve less capital than equivalent increases in energy supplies. The real problem is not so much the amount of capital needed to make efficiency improvements; it is much more the institutional and organizational hurdles that stand in the way.

Consider a decision to spend $2 billion on either a centralized power plant or end-use efficiency improvements. Constructing a large power plant is a relatively straightforward task that can be carried out by a small and disciplined team. Spending $2 billion on end-use improvements is much more complicated. If each efficiency measure costs between $2,000 and $20 million, then between 100 and a million subprojects are involved. Organizing so many diverse activities is difficult.

Marketing end-use improvements is inherently more complicated than marketing conventional energy supplies or end-use devices. It is not enough to be concerned just with the "hardware" of the new energy-efficient devices; equal attention should be paid to the essential "software" that will help customers use and understand the benefits of the devices. All aspects of the marketing of conservation should be addressed if the aim is to promote energy end-use efficiency in developing countries.

Electric and gas utilities are particularly well qualified to market end-use efficiencies. Already some of the more progressive utilities in the U.S. have begun to sponsor conservation programs for consumers (see Chapter 2, "Efficient Use of Electricity," by Arnold P. Fickett, Clark W. Gellings and Amory B. Lovins). Utilities are also well positioned to direct their considerable capital resources to energy-efficiency investments. They have an administrative structure for channeling capital to the many households and business consumers with which they deal. Moreover, their regular billing system is an ideal vehicle for permitting customers to repay loans from the utility.

Energy conservation can be marketed comprehensively in various ways. One possibility is to convert energy utilities, which merely supply energy, into energy-service companies that provide heating, cooling, lighting and other services. Financial institutions can help motivate utilities to change themselves into energy-service vendors.

For any of these improvements to occur in developing countries, the local, state and federal governments must learn to formulate energy policies in new ways. Above all, the level of energy services, rather than the magnitude of consumption, must be considered the true energy indicator of development. Highest priority should be given to extending energy services that improve the quality of life for the poor, generate employment and influence critical economic sectors, such as agriculture and industry.

As we have seen, planner scan make energy services more available either by producing more energy or by improving the efficiency of devices that deliver the services. The costs and environmental impacts of the competing solutions must be compared fairly to determine which approach is better. Energy saved through greater efficiency should be treated like additional energy that has been produced. Efficiency measures and decentralized power plants should be compared with centralized plants according to the same terms of credit and interest rates, benefits, incentives, subsidies and other considerations.

At present the competition is unfairly weighted in favor of centralized sources and against efficiency measures. Some specific policies should be spelled out for promoting fair competition through the elimination of subsidies for energy supplies, pricing that takes into account long-run marginal costs and the generation of sound data bases for the comparison.

Policy instruments available for the implementation of energy policies include market forces, subsidies, taxes, regulated standards and labeling practices. Each instrument has a special degree of effectiveness — and ineffectiveness. The implementation of any efficiency measure will derive from a specific package of policy instruments.

The market is an excellent allocator of capital, raw materials and manpower, and its strengths must be fully exploited. But market forces have their limits. In particular, they cannot be trusted to safeguard social equity, long-term interests and the environment. Special policies must be devised to protect the poor, the environment and other concerns.

If international aid agencies change their practices, they, too, can help implement DEFENDUS strategies in developing countries. When allocating aid, the agencies must compare energy-supply increases and efficiency improvements on the same terms as part of a single decision-making process rather than considering each separately. The agencies should fund conservation and renewable-energy programs. And when evaluating various projects, the agencies should consider the global environmental impact.

Moreover, financial-aid programs should be restructured to shift their emphasis from support for specific projects (such as building dams) to support for goal-oriented programs (such as lighting more homes). Aid agencies should build and strengthen the institutions in developing countries, boost the countries' energy-related technical capabilities and support technological leapfrogging.

National and multinational companies can play a crucial role in promoting technological leapfrogging. Through collaborative ventures with local manufacturing firms, the companies can transfer the latest energy-efficient technologies to developing countries.

If the developing world has access to funds to implement DEFENDUS energy strategies, the global environment will benefit tremendously. The money will have to come from the industrialized world, however. The funds could be collected through a tax that is proportional to a country's contribution to the total carbon dioxide emissions. Such a carbon tax would be consistent with the "polluter pays" principle that was accepted long ago (at least in theory) by the nations in the Organization for Economic Cooperation and Development (OECD).

As capital becomes scarcer and environmental concerns grow, it will become more and more difficult for funding institutions to reject quantitative proposals for cost-effective, environmentally sound energy plans. Nevertheless, shifting from a supply-biased approach to a more balanced one will demand fundamental changes at many levels. Furthermore, decisions about energy are not always made rationally — powerful vested interests have grown up around the conventional energy-supply industries.

Can the needed transformation in energy policies take place in the next few decades? Aside from actions by international aid agencies, the best hope for change lies in a convergence of interests. Industrialized countries are now threatened by the global environmental consequences of conventional energy strategies in developing countries. They are beginning to realize that supporting sustainable development is in their own best interest. The growing environmental movements in the developing countries are forging alliances with their counterparts in the industrialized world. Prodevelopment activists are speaking out against supply-biased strategies because the additional energy supply is not trickling down to the poor. Not surprisingly, the poor are becoming restless with the energy systems that have failed them.

The lesson is clear: DEFENDUS scenarios may be difficult to implement, but the current energy systems are impossible to sustain.

Energy for the Soviet Union, Eastern Europe and China

*Economic reforms and new technology may allow the centrally
planned economies and the emerging democracies to
develop without further harm to the environment.*

• • •

William U. Chandler, Alexei A. Makarov and Zhou Dadi

Thomas Jefferson described resolution as the extraordinary event necessary to enable all the ordinary events to continue. The extraordinary events of 1989 in Eastern Europe, like Chinese economic reforms a decade ago and Soviet *perestroika*, were born of necessity to improve material well-being in the socialist countries. Yet the stark images of energy-related smog and soot that have emerged with the new openness warn that these economies cannot improve themselves at the expense of the environment (see Figure 7.1). As in the market economies, an energy revolution is needed.

A revealing statistic points toward a solution. The centrally planned economies of the Soviet Union, Eastern Europe and China support one-third of the world's people, and they consume one-third of the world's energy to produce one-fifth of the world's economic output. Energy intensity — the amount of energy it takes to produce something, a ton of steel, say — is strikingly higher in the planned economies than it is in market economies (see Figure 7.2).

How can the planned economies reduce energy intensity while providing more living space, consumer goods and higher incomes? Will attempts at restructuring be translated into wider energy use? If these economies grow rapidly (as China's has begun to do), can the environmental impact be moderated? In this chapter we take a close look at these questions and ask how economic reform, technological change and international cooperation can help the Soviet Union, Eastern Europe and China achieve both economic development and environmental protection.

The high energy intensity of planned economies does not reflect a profligate way of life. It is the result, in part, of outmoded technology and excessive investment in heavy industry. Although energy consumption per capita in Eastern Europe and the Soviet union matches that of much richer Western Europe, the energy is used in a different way — to produce energy-intensive steel, aluminum and con-

Figure 7.1 ENVIRONMENTAL CONTROLS and economic development in Eastern Europe lag behind the West by several decades, Energy use accounts for the heavy burden of airborne particulates in this scene from the Hesse region of East Germany.

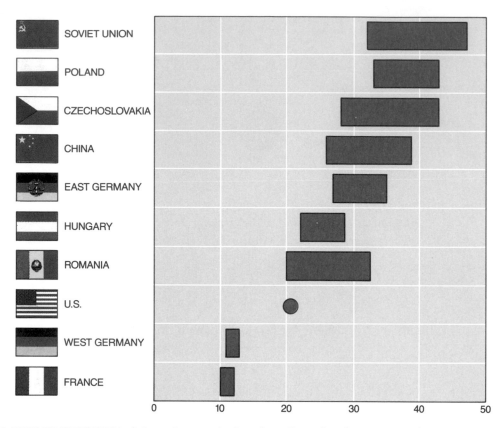

Figure 7.2 ENERGY INTENSITY of planned economies is difficult to compare with that of market economies because of different ways of measuring economic output and because of fluctuating exchange rates. As indicated by the bars, the authors have attempted to compare energy intensity using a range of estimates of economic output corrected for purchasing power and currently fluctuations.

crete rather than to power automobiles or run home appliances. The U.S., for example, has enough cars to put every American in an automobile at the same moment, with no one having to sit in the back seat; the Soviet Union has one car for every 20 persons; China provides only one car for every 1,000 people. The Soviet Union, on the other hand, uses 30 gigajoules (30 billion joules; one billion joules is the equivalent of about one million Btu's) to produce a ton of steel, compared with 18 gigajoules to produce a ton in Japan.

I n addition to their automobiles, the citizens of most developed countries enjoy more living space than their counterparts in the planned economies: 30 square meters for Western Europeans and 55 for Americans, compared with 20 square meters or less for Soviets and Eastern Europeans—and 6.5 for the average Chinese (see Figure 7.3). More living area usually means more energy is consumed for heating and for operating appliances. Energy use per square meter of living space in Eastern Europe, however, is from 25 to 50 percent higher than it is in the U.S., reflecting the lack of individual heating controls in apartments and the absence of economic incentives to conserve heat and hot water.

Chinese households face altogether different problems. In many provinces, indoor winter temperatures fall below freezing, but the government, which strictly controls the residential use of coal, provides no energy for heating houses. Air-conditioning is nonexistent in households, although indoor temperatures in south and middle China often exceed 35 degrees Celsius (95 degrees Fahrenheit) even at midnight.

Figure 7.3 AMOUNT OF FLOOR SPACE in Soviet (*top*) and Chinese (*bottom*) homes is only one half and one fourth, respectively, that of French homes. Whereas Soviets enjoy the basic amenities of modern life — heat, hot water, refrigeration, television — many Chinese have no space heating, no refrigerators and only cold water. China will be unable to meet the energy demands of such amenities without obtaining the most efficient appliances, buildings and energy systems available in the world.

Currently a Chinese household consumes 36 times less energy than an American one does. Average electricity consumption per household is less than 120 kilowatt-hours a year. A relatively efficient American refrigerator uses that amount in two months. Fewer than 10 percent of all Chinese households have even a small refrigerator. Many rural areas have no electricity or gas: firewood and crop stalks are burned in primitive stoves for cooking and heating. Most Chinese households still burn coal; seven-eighths of the energy for residences is supplied by that high-sulfur, inefficient fuel.

The differences in energy use between East and West have affected more than the standard of living. In the planned economies, air, water and human health are in jeopardy.

In the Soviet Union, for example, energy production had, by 1989, increased concentrations of toxic air pollutants to a level 10 times the maximum permissible in 88 Soviet cities with a combined population of 42 million people. Reservoirs for hydroelectric projects cover one tenth of the area suitable for urban development and industrial and transportation facilities. Twenty large hydroelectric-power stations erected on lowland rivers have transformed flowing rivers into stagnant reservoirs and reduced the ability of these waters to assimilate wastes. The Chernobyl reactor accident killed several dozen people and contaminated 10,000 square kilometers with levels of radioactivity that exceeded 15 curies per square kilometer, affecting more than 250,000 people.

Environmental problems in Eastern Europe — many of them directly related to energy production — have reached crisis proportions. In Poland and Czechoslovakia, sulfur dioxide deposition averages

12,000 micrograms per square meter every month, from four to eight times that in most European countries. Toxic elements, including lead and arsenic released from power plants that burn high-sulfur coal, accumulate in the soil and contaminate food. In the worst-affected communities of Czechoslovakia, bone growth in one-third of the children is retarded by 10 months or more.

In China, coal combustion pollutes both air and water. Suspended particulates present a major health hazard in all cities of northern China, where annual concentrations average 740 micrograms per cubic meter. In contrast, the yearly average in much of the U.S. is 50 micrograms per cubic meter. Sulfur emissions in China now total more than 15 million tons a year, bringing the acidity of rain to a pH level below 5 over an area of 1.3 million square kilometers. Because of low pH soils and high humidity, the problem is particularly severe south of the Yangzi River, where the pH of rain in some areas measures below 4.5. Meteorologists at the Chinese Environmental Protection Agency estimate that in Guangdong province alone, in the southernmost part of China, economic losses caused by acid rain amount to as much as .1 percent of gross domestic product (GDP). The economic loss caused by all forms of pollution in China is estimated to be about 2 percent of the GDP.

Few coal-burning facilities in planned economies are equipped with desulfurization equipment, and only medium-to-large industrial boilers have devices to control particulates. Most burners emit smoke without any controls; coal is rarely washed to reduce its sulfur content. Expensive pollution-control equipment has not been a priority for planners and, against the background of economic difficulties throughout the planned economies, will not likely become one in the 1990's.

How can planned economies protect their natural environments while meeting growing economic expectations? Price reform, a shift to market mechanisms and energy-efficiency laws and standards will play crucial roles. Price reform is a proved way to promote energy-saving behavior. When the U.S. decontrolled the price of energy at the end of the 1970's, it sent an unmistakable signal to consumers that saving energy was important. As a result, the U.S. increased economic growth by 40 percent between 1973 and 1988 while essentially holding energy demand constant. Up to two-thirds of that achievement was prompted by consumer response to higher prices. Prices in the Soviet Union must be increased by a factor of almost three. Yet it is not enough simply to get the prices right: prices must also matter. Making prices matter means not permitting enterprises to pass the cost of energy waste on to consumers. This change means shifting to market mechanisms to promote efficiency through competition.

Market mechanisms alone, however, will not complete the energy revolution in the planned economies. All market economies still grapple with energy inefficiency caused by market imperfections such as lack of information, monopoly control of electric-power production and shortage of financing—not to mention the extravagances of wealth. For these reasons, some market economies impose minimum performance standards for energy-consuming equipment such as automobiles and furnaces. The Soviet Union, Eastern Europe and China can combine price, structural and regulatory reform to alter the present situation.

In China at present, as in all planned economies, governmental agencies control the energy market through the allocation of capital and other resources among sectors. In addition, local governments directly manipulate economic activity and hamper the transfer of resources between regions. A multitiered system of prices, for example permits mines to sell coal that exceeds planned quotas. "Out of plan" coal may be sold to enterprises that want to manufacture other goods out of plan, which can then be sold for the profit of employees.

The price of goods, particularly coal, produced this way can rise above international levels. Prices can also be distorted by local governments, which control the transfer of coal across regional boundaries. Worse, a chronic shortage of rail capacity limits such transactions and raises their cost even higher. The shortage of coal forces many factories to shut down two or three days a week and many enterprises resort to paying large fees to operators of small, gasoline-powered trucks to haul coal hundreds of kilometers.

The potential for any economy to meet energy and environmental goals can be best assessed by making projections of energy demand and the resulting emissions. Analysts in planned economies have recently developed models to generate scenarios that help clarify where structural change and new technology will lead (see Figure 7.4).

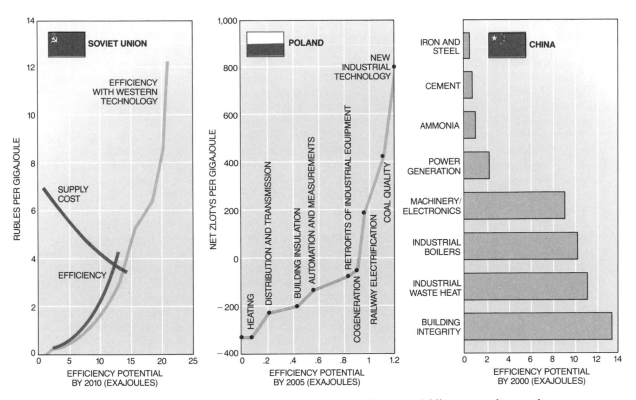

Figure 7.4 POTENTIAL FOR ENERGY EFFICIENCY in the Soviet Union compares cost of efficiency options with cost of additional production of energy resources. The Energy Research Institute of the Soviet Union provided data. Energy efficiency in Poland is shown in terms of net cost; alternative energy-supply cost has been subtracted from cost of efficiency, yielding a negative cost in many cases. Data are from S. Sitnícki of the Polish Environment Ministry. Estimated potential for China assumes use of the best technology available on international markets. China's Energy Research Institute furnished data.

Two scenarios for the Soviet Union, created by one of us (Makarov) and his colleagues from the Soviet Energy Research Institute, illustrate possible avenues of economic development. In the first scenario the transformation to a market economy takes place over a 10-year period and is only partially successful. As a result, the economy grows at a rate of 2 to 2.2 percent annually through the year 2000, increasing thereafter to 2.5 to 2.8 percent. Analysts believe this level of growth is the minimum necessary to avert destruction of Soviet society.

The second scenario is more optimistic: it assumes that rapid economic reforms, which include a move to market pricing and the competitive production of goods and services, will succeed within the next four or five years. These changes would spur growth at an annual rate of 3 percent between 1991 and 1995 and accelerate to 3.8 percent in the following years.

In the pessimistic scenario, living space per capita would not reach current Western European levels of 30 square meters until 2030. The second scenario would achieve this level by 2015. Consumer goods would triple by 2030 in the pessimistic case, by 2010 in the optimistic one. The Soviet Union would not reach the Western European level of cars per capita until after 2030 in the pessimistic case but would do so in the optimistic case by 2025.

Restructuring the Soviet economy has major implications for energy conservation. Materials use per capita there is high compared with other nations, and market reforms would have two effects: less unnecessary material would be produced, and materials would be used more efficiently in manufacturing, construction and packaging as manufacturers compete for price-sensitive markets. Structural reform could, in the optimistic scenario, reduce the Soviet demand for energy by one-sixth over the next two decades.

The Soviet Union could save still more energy by applying its own state-of-the-art technology. Priority Western imports include efficient lighting systems, power plants, industrial ovens and the capability to control industrial motors electronically to match motor speed to the speed needed to do the work. Introducing the best technology available over the next 20 years would reduce future demand by one-third.

Soviet economists are concerned that excessive investment in energy production is reducing economic growth. Annual investment in oil, natural gas and coal production has reached almost 20 percent of total annual investment. The goal is to reduce this share about 17 percent in the next five years and to 13 percent by early in the 21st century.

Energy efficiency is crucial for *perestroika*: it frees scarce domestic resources such as capital and technical skills for use in modernizing industry and agriculture while freeing oil and gas for export to earn hard currency.

A scenario for development in the countries of Eastern Europe assumes a growth rate of 3 percent a year. This rate represents an optimistic future in which planners somehow find a way to maintain past economic growth without radically changing their economies. Current trends would increase energy demand by almost 50 percent (from 20 to 31 exajoules) by 2025. (An exajoule, 10^{18} joules, is equivalent to one quadrillion Btu's.) But, as in the Soviet Union, a combination of economic reform and the introduction of energy-efficient technology could hold energy demand virtually constant.

Reform would close down factories that because of weak price signals yield a high proportion of scrap or unneeded output. In Czechoslovakia, for example, steel production could probably be cut to half the present level of almost 16 million tons a year by reducing waste. Even more dramatic changes have been recommended for the production of nonferrous metals and chemicals. Structural change will reduce the role for heavy industry in national economic output, with increases coming in the manufacture of consumer goods and in services.

Poland is the largest energy producer and consumer in Eastern Europe, accounting for one third of the 20 exajoules Eastern Europeans use each year. Until recently; electricity, coal and natural gas were priced at one-fourth, one-half and four-fifths of world market levels. Without economic reform, the Polish economy would double its energy consumption by 2025. A combination of reform and energy efficiency, in contrast, would hold energy demand steady during that period.

In the industrial sector, energy-efficient motors, process automation and cogeneration (simultaneously generating electricity and heat for industrial processes or use in water or space heating) would offer large savings. At the same time, metering and individual control of residential heating and hot-water systems, coupled with repairs and improved management of district heating systems, could permit housing space to double with only 15 percent growth in energy consumption.

Poland is the world's fourth largest coal producer and obtains three fourths of its domestic energy from coal. In addition to severe problems of acid rain and suspended particulates, coal mining imposes heavy supply burdens on the economy. It consumes one-fifth of all steel used in the country, for mine roof and structural supports and nearly one-tenth of all electric power. Coal production becomes more difficult and expensive year by year because the resource is being depleted. The average depth of mines is now 600 meters, and every year the depth of new mines increases 10 to 20 meters. Reducing the demand for coal would simultaneously free capital for more productive uses elsewhere in the economy and reduce the burden of particulate, sulfur and toxic emissions from coal combustion.

Eastern European incomes will probably grow to match Western levels by 2025, as will living space per capita. Larger refrigerators, more washers and dryers, air conditioners and personal appliances will mean higher energy intensities despite improvements in water and space heating. In the absence of policies, energy demand for the residential sector would more than double by 2025 (from 5.2 to 12.5 exajoules). Realistic energy pricing, metering use, and efficiency standards for appliances similar to those in the U.S. could hold demand to 7.6 exajoules.

As incomes grow in Eastern Europe so will the demand for transportation. With the inefficient automobiles in current use, transportation energy demand would double—and would be very costly, because it would almost certainly require an increase in oil imports of a million barrels a day. Cars in Eastern Europe today average 8.7 liters per 100 kilometers (about 27 miles per gallon), which is very inefficient given the low weight and power of the cars. Yet increasing automobile fuel economy to five liters per 100 kilometers (about 47 miles per gallon) and matching truck size and power to loads can help hold transportation energy demand to 3.4 exajoules, which is a 50 percent increase over 1985 levels.

Few countries use energy less efficiently than China. In 1988 China consumed at least three times more energy for each unit of economic output than Japan did. Moreover, China's intensive use of energy cannot be attributed to its "stage of development," for other developing countries, such as South Korea, derive similar shares of national income from heavy, energy-intensive industry while using much less energy. Detailed studies by the

World Bank show that energy intensity in steel production is twice as high in China as in Japan and that of truck transport is twice as high as in the U.S. (see Figure 7.5).

China has nonetheless made impressive progress, cutting energy required per unit of economic output by 4.7 percent annually over the past decade. It is true that energy intensity was so high in the first place that reductions were easy, but the achievement remains unmatched by any other developing nation. In 1980 Chinese leaders consciously set out to achieve this goal, because they recognized that they could not physically supply the amount of energy needed if they wanted to realize their ambitious goals for economic growth. Economic growth in China—exceedingly difficult to estimate in Western terms—may have averaged 10 percent a year over the past decade, while coal use grew only half as fast. And coal supplies three fourths of China's energy.

A plausible scenario for the Chinese economy

Figure 7.5 TRANSPORTATION STOCK in China is small —fewer than five million vehicles—and inefficient. Trucks often burn gasoline instead of more-efficient diesel fuel and make many empty return trips because of distortions in the planning system.

would call for continued economic growth at a rate of about 5 percent of the next three decades. Even this quadrupling of per capita wealth would provide income levels in 2025 only one-third those in the U.S. today. The population will grow to at least 1.4 billion unless the policy of one child per family succeeds — and it appears to be failing in rural areas. The number of automobiles would likely climb to one car for every 50 persons. Residential energy demand would increase almost 50 percent. Overall, Chinese energy demand would nearly triple (from 33 to 86 exajoules) by 2025.

The increase in demand presents a formidable supply problem. Oil consumption would grow from two million to almost seven million barrels a day. Coal consumption would triple. Even with price reform, economic restructuring and modest technical efficiency standards, total energy use would grow to 75 exajoules, close to current U.S. levels, and most of it would be supplied by coal. Carbon dioxide emissions would grow from the current 580 million tons a year to 1.5 billion, for an increase of global carbon emissions equal to 17 percent of today's level. Making economic and environmental goals converge in so poor a country as China will be a challenge. The target for carbon-emissions reduction, for example, must be not to cut emissions but to hold growth to a minimum. The need for growth in developing countries thus increases the need to cut emissions in developed countries even more.

Each of these scenarios raises new questions. How will the increased demand for oil be met? What part will coal, which threatens the global ecosystem, play? What about nuclear power? Can natural gas fill the gap? If so, where will it come from?

Oil remains a question mark in the planned economies. The Soviet Union, by far the world's largest oil producer, supplying one-fifth of world production, could possibly maintain its current level of output until 2010. This estimate is based on increased world oil prices and the adoption of Western technology to attain maximum output. The small amounts of petroleum produced in Eastern Europe are expected to continue to decline.

China currently ranks sixth in the world in oil production at 2.7 million barrels a day, and this rate could be sustained for about 13 years with proved reserves. Oil-bearing sedimentary basins in China cover some 5.5 million square kilometers and theoretically contain some 550 billion barrels of oil. But oil exploration and development are constrained by lack of modern equipment and capital.

Moderate growth of coal production is envisioned for the Soviet Union, with output increasing from about 13 exajoules in 1990 to 17 or less in 2010. The future of coal is more clouded in Eastern Europe, where economic and environmental problems make its use highly undesirable. Coal reserves in China would, at current rates of consumption, last 1,000 years. Even this prodigious quantity would not require miners to go deeper than 1,500 meters.

The Soviet Union expects only moderate expansion of its nuclear-power industry, which may reach from 45 to 65 gigawatts by the year 2000 and from 60 to 100 by 1010. Even this prospect is contingent on the rapid development of safe reactors at reasonable cost. Nuclear power has become controversial in Poland, Czechoslovakia and Hungary. Poland has no reactors in operation and is not likely to complete those that are under construction. China possesses the capability to produce reactors up to 600 megawatts in size, but the large investments of capital and foreign exchange necessary — and the enormous reserves of low-cost coal — make large-scale development of nuclear power unlikely.

The Soviet Union's immense natural gas reserves offer an attractive energy strategy — a gas bridge to the future. The Soviet Union envisages growth in gas production from 29 exajoules in 1988 to as much as 35 by the year 2000 and 38 by 2010. This possibility opens up many others. One promising option is the fundamental reorientation of the utility sector. Natural gas supplies could be stretched significantly by adopting highly efficient combined-cycle systems that produce both electricity and steam for residential heating or industrial processes. The technology would reduce the environmental problems in the Soviet Union associated with coal and nuclear power and would conserve gas for export.

The prospect of greater reliance on Soviet gas in Eastern Europe is very attractive. Three conditions must be met, however, before the prospect can become a reality. First, economic reform is required to avoid wasting the resource. Second, economic recovery is necessary to enable Eastern Europe to earn foreign exchange to pay for Soviet gas. Indeed, the net effect of setting the import prices for Soviet oil and gas and the export prices of Eastern European products at world market levels would cost the region at least $10 billion a year. Third, highly efficient technology for combined-cycle gas and steam turbines must be introduced to Eastern Europe. This technology would help stretch the resource and keep costs affordable. Again, however, the problem of foreign exchange will impede acquisition of the technology.

One solution to the foreign exchange problem would be a creative joint venture among Soviet and Eastern European enterprises, Western gas-steam turbine manufacturers and Soviet gas producers to finance the production of gas-using equipment and the delivery of gas to Eastern Europe consumers. The Eastern Europeans would gain a cleaner, more productive energy system, the Soviets would gain a

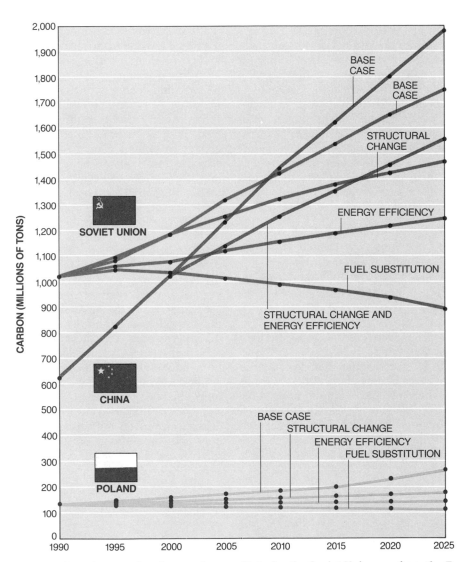

Figure 7.6 COMBINATION of economic reform and energy efficiency offers the greatest opportunities for reducing the growth of carbon dioxide emissions. In a country as poor as China, however, the target must be not to cut emissions but to try to hold their growth to a minimum. Data for the Soviet Union are from the Energy Research Institute; S. Sitnícki provided estimates for Poland; Chinese projections were done by a team led by J. Sathaye at the Lawrence Berkeley Laboratory.

market for their gas and the Western counties would gain export markets. The entire world would benefit from reduced carbon and sulfur emissions.

Major quantities of natural gas could be produced in East Siberia and the Okhotsk Sea shelf, some of which could be exported to China, Korea and Japan. The Soviet Union and Asia have recently considered the export of gas to Korea through northeastern China. Gas fields near Yakustk or off Sakhalin Island could supply from two to three exajoules a year to China, or about 10 percent of the country's current energy needs. Constraints on this arrangement include foreign exchange for China to pay for the gas and foreign technology for the Sakhalin fields (because offshore platforms are required).

The Soviet Union could completely cover its domestic energy needs through the exploitation of high-grade resources. Oil and gas would thus continue to represent about 80 percent of total Soviet energy supply. According to the optimistic case of economic reform, total energy exports to Europe would slightly decrease; declining oil shipments would be almost offset by increasing transmission of natural gas.

In the above scenarios, the growth of greenhouse-gas emissions in the Soviet Union could be reduced by 255 million tons in 2010, about one fourth of current emissions. Similarly, toxic gas emissions could be cut by 30 to 40 percent of present-day levels (see Figure 7.6).

Opportunities for reducing carbon emissions exist also in Eastern Europe. If Poland, for example, chose to do so, it could cut carbon dioxide emissions by 20 percent at a cost of only about .3 percent of its GDP by 2010. The possibility hinges in part on the availability and affordability of Soviet natural gas. Poland will not be able to solve its energy and economic problems without obtaining Western technologies. Borrowing money to import products

or process technology, however, seems unlikely given Poland's burden of foreign debt. Joint ventures with Western corporations could be one important means of technology transfer, for they provide both financing and technical expertise. It appears that Polish leaders understand the importance of getting energy price signals right, and price reform and competition will greatly reduce the demand for more energy.

If the international community actively participates in transferring energy conservation technology more impressive results will be obtained. Western governments can aid this effort financially by providing loan guarantees, credits or even direct financing. They could establish energy conservation funds within existing institutions or new ones created for that purpose. Such funds, however, will always be limited. Joint ventures with the private sector in the West can help. But internal resources must be brought to bear by shifting large sums of money from military purposes and from traditional concentration on energy supply. Indeed, Stanislaw Sitnícki, chief adviser to the Environment Minister of Poland, argues that his nation should simply suspend new investments in supply until it gets the demand side of its house in order.

Italian socialist writer Antonio Gramsci wrote 60 years ago, in a very different context, in *Prison Notebooks* that "the crisis consists precisely in the fact that the old is dying and the new cannot be born; in this interregnum a great variety of morbid symptoms appear." Most socialist nations seem to be headed for a new system of greater market reliance, but because no nation has ever made the transition from a planned to a market economy, much remains to be learned. With regard to energy, success will depend in part on the cooperation of Western nations and corporations. But the tough issues can be solved only from within.

Energy from Fossil Fuels

*Until other energy sources supplant coal, oil and natural gas,
the technological challenge is clear: extract maximum energy
from the old standbys while minimizing harm to the environment.*

. . .

William Fulkerson, Roddie R. Judkins and Manoj K. Sanghvi

Fossil fuels pose a dilemma for human society. Worldwide, the combustion of coal, oil and natural gas supplies some 88 percent of the energy we purchase and makes much of what we do possible. Yet gases emitted during burning can degrade the environment, perhaps to the extent of altering the climate and threatening the future habitability of the planet.

Technologies now being developed should go a long way toward alleviating two regional-scale environmental disturbances associated with fossil-fuel combustion: acid deposition and urban smog. Solutions are also being sought for a third and potentially more devastating disturbance—increased global warming caused by rising levels of carbon dioxide (CO_2) and other so-called greenhouse gases in the atmosphere. CO_2 is emitted whenever fossil fuels are burned (see Figure 8.1). Like other greenhouse gases, it captures heat radiated from the earth and traps it near the surface.

The problem of global warming is difficult to resolve, in part because there is disagreement over how real and dangerous the threat is and, consequently, how much money and effort should be devoted to coping with it. Still, we think uncertainty should not be an excuse for inaction.

Society needs to take out some technological insurance against the possibility that the world will have to curtail fossil-fuel consumption drastically and rapidly to prevent global climate change. A number of relatively inexpensive strategies worth considering are beginning to emerge. These range from increasing the efficiency of fossil-fuel use (which would reduce emissions by producing more energy services with the same amount of fuel) to developing improved nonfossil sources that can substitute economically on a large scale.

If environmental stress does not force a shift to other fuels, the finite supply of fossil fuels will mandate such a change eventually (perhaps in a century or two). The fossil supplies were laid down over the millennia, as ancient plants and animals died and became buried in swamps, lakes and seabeds. Much fuel remains, especially coal. Nevertheless, stockpiles are being depleted many times faster than they are being replenished.

There are good reasons why fossil fuels are so popular. First, they are accessible in one form or another in all regions of the world. Second, humankind has learned to exploit them efficiently and relatively cleanly to produce the energy services it needs. The relatively simple technology of controlled combustion provides energy for applications

at nearly every scale. Third, they make excellent fuels for transportation—because they are portable and store a great deal of chemical energy and because the oxygen required for combustion is ubiquitous in the air. Finally, one form can readily be converted to another, say, from solid to liquid or gas, and the fuels are excellent feedstocks for chemicals and plastics.

For these reasons, and because it will probably take many decades before competitive alternatives are developed and widely adopted, fossil fuels will continue to be important for quite some time (see Figure 8.2). Therefore, technologies to ameliorate the effects of fossil-fuel combustion on acid deposition, urban air pollution and global warming must be pursued. (Of course, the environment is not the only important issue relating to fossil fuels. The stability of oil markets and energy security are also major concerns, although we do not have enough space to discuss them.)

The solution to the first environmental problem—acid deposition—lies to a great extent in controlling the release of sulfur dioxide (SO_2) and oxides of nitrogen (NO_x), which are converted to acids when they combine with water in the atmosphere. Combustion of fossil fuels, especially coal, accounts for more than 80 percent of the SO_2 and most of the NO_x injected into the atmosphere by human activity. The nitrogen comes from the fuels and from the air, and it combines with oxygen to form NO_x when combustion temperatures are high.

Whatever succeeds in reducing the output of NO_x should also help reduce urban smog, a product of sunlight acting on NO_x, carbon monoxide (CO) and certain organic compounds. Much urban pollution is caused by vehicles, which run almost exclusively on petroleum products. (Technologies for reducing emissions from vehicles are addressed in Chapter 5, "Energy for Motor Vehicles," by Deborah L. Bleviss and Peter Walzer.) We should note, however, that a great deal has been done to reduce vehicular emissions, and proposed revisions of the U.S. Clean Air Act make further reductions probable. Indeed, the mandate to improve urban air quality may stimulate large-scale experiments with alternative fuels and

engines and even the transportation system itself. Such experiments may yield new insights into ways to control emissions of greenhouse gases by the transportation sector.

Because more than half of the SO_2 and some 30 percent of the NO_x released into the atmosphere by human endeavors come from the combustion of coal, a number of the technologies aimed at diminishing the output of these gases have been designed for coal-fired electric-power plants. These plants use about 85 percent of the coal burned in the U.S. The development of pollution abatement technologies is being greatly facilitated in the U.S. by the Clean Coal Technology Program, an ambitious effort funded jointly by government and industry.

In conventional power plants, pulverized coal is burned in a boiler, where the heat vaporizes water in steam tubes. The resulting steam turns the blades of a turbine, and the mechanical energy of the turbine is converted to electricity by a generator. Waste gases produced in the boiler during combustion—among them, SO_2, NO_x and CO_2—flow from the boiler to a particulate removal device and then to the stack and the air. Typical plants run at an efficiency of 37 percent.

Today the method in widest application for reducing SO_2 emissions is flue-gas scrubbing. In scrubbers, which can be added to existing equipment, exhaust gases come in contact with a limestone ($CaCO_3$) slurry or some other sorbent and react to form a compound that can be removed as solid waste. As a result of scrubbing and an increased reliance on low-sulfur coals, SO_2 emissions have decreased by 33 percent during the past 15 years in the U.S., even though coal use has increased about 50 percent. NO_x emissions have remained relatively constant. Standard scrubbers can shrink SO_2 emissions by 50 to 90 percent, albeit at some cost to efficiency. The typical steam plant outfitted with scrubbers runs at an efficiency of about 34 percent. About a third of the chemically stored energy of coal is converted to electricity, and the remaining two thirds is lost or diverted to run the plant's machinery.

Certain advanced scrubbers now being developed can remove as much as 97 percent of the SO_2 and, depending on the sorbent used, may also reduce the release of NO_x. After combustion, NO_x can be removed by selective catalytic reduction; ammonia (NH_3) is mixed with exhaust gases, where it reacts with the NO_x to form benign substances, namely, water and molecular nitrogen (N_2), a major constituent of air.

Figure 8.1 WORM'S-EYE VIEW is of a natural gas pipeline in Texas. Where the gas is abundant, it might be substituted economically for coal. Natural gas burns more efficiently and cleanly and yields only half as much carbon dioxide (CO_2) for each unit of energy produced.

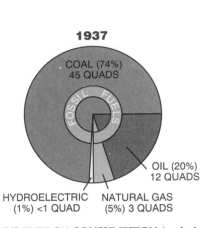

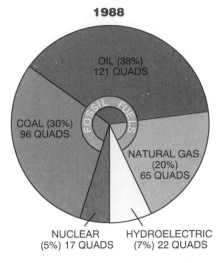

Figure 8.2 WORLD'S ENERGY CONSUMPTION (excluding that from biomass) rose from 60 to 321 quads between 1937 and 1988. In that time, nuclear power was introduced, and waterpower came to account for more of the energy pie, but worldwide dependence on fossil fuels continued.

Even before combustion begins, coal can be cleansed of some sulfur. For instance, commercially available processing methods crush the coal and separate the resulting particles on the basis of density, thereby removing up to about 30 percent of the sulfur. Investigators are studying more effective mechanical approaches, as well as new chemical methods and even biological ones.

Techniques that require only retrofitting can also eliminate some SO_2 or NO_x during the combustion process. Sulfur, for example, can be removed by injecting limestone along with coal into a boiler. Emissions of NO_x can be controlled by several methods that lower combustion temperatures, such as injection of steam into the combustion region.

As new technology is developed, unwanted emissions may be reduced by repowering: replacing aging equipment with more advanced and efficient substitutes. Such repowering can be more economical than retrofitting. An old power-generating system might, for example, be exchanged for one based on not a conventional steam boiler but a new kind of combustion chamber—such as the atmospheric fluidized-bed combustor (AFBC) or the pressurized fluidized-bed combustor (PEBC).

Both systems, which are now in limited commercial use by utilities and industry, burn coal with limestone or dolomite in a mixture suspended by jets of air. The mixture, or fluidized bed, and the hot combustion gases envelop clusters of steam-generator tubes in and above the fluidized bed. The limestone sorbent takes up about 90 percent of the sulfur that would normally be emitted as SO_2. The constant churning facilitates both the burning of the coal and the transfer of heat to the tubes. In these systems the temperature of combustion can be lower than in a conventional boiler, thereby reducing the formation of NO_x.

In a major advance over the atmospheric version, the pressurized variety maintains pressure in the boiler that is six to 16 times greater than standard atmospheric pressure. Extra efficiency is achieved by exploiting the hot gases in the combustion chamber in what is called a combined cycle. As the vapor in the steam-generator tubes drives a standard steam turbine, the pressurized gases run a gas-driven turbine.

An important repowering approach attracting great interest is the integrated coal-gasification combined cycle (IGCC) system, now being demonstrated at several sites around the world (see Figure 8.3). The major innovation introduced with this technology is the transformation of coal into synthesis gas, a mixture of mainly hydrogen (H_2) and CO, with lesser quantities of methane (CH_4), CO_2 and hydrogen sulfide (H_2S). Up to 99 percent of the H_2S is removed by commercially available processes before the gas is burned.

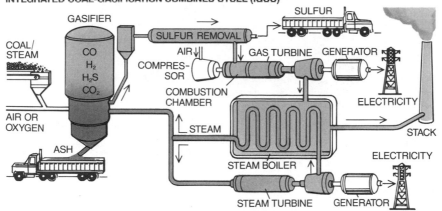

INTEGRATED COAL-GASIFICATION COMBINED CYCLE (IGCC)

Figure 8.3 ONE ADVANCED TECHNOLOGY for generating electricity—IGCC—converts coal to a gaseous mixture (*yellow*) by reaction with steam and oxygen before burning; the system then runs what is called a combined cycle. The hot gases power a gas-driven turbine, and the heat from the gases is recovered to vaporize water (*blue*), which runs a steam turbine. Combined cycles are more efficient than conventional steam cycles because they extract more energy per unit of coal burned. In conventional power plants, coal is burned directly in a boiler to run a steam turbine. The gases produced during the burning are not used but are released through a stack or diverted to a sulfur-removing scrubber before release.

The synthesis gas then powers a combined cycle something like that described above. In current designs the hot gases are burned in a combustion chamber to drive a gas turbine. Then the exhaust gases from the turbine generate steam to drive a steam turbine. Efficiency claims for the pressurized fluidized-bed and integrated coal-gasification combined cycle systems are similar—about 42 percent.

The latter system and other coal-gasification approaches are particularly important to the future of coal as an energy source because they make great flexibility possible. Gas from coal not only can run combined gas and steam turbine cycles but also can be exploited for cogeneration, in which steam is tapped for heating applications. Moreover, the gas can substitute for natural gas and serve as a source of hydrogen to power fuel cells.

Because of their higher efficiency, advanced repowering technologies can help limit the more copious, and more troubling, emissions of CO_2, which is a major and long-lived greenhouse gas and accounts for about half of the volume of such gases in the atmosphere. It may be that human-generated emissions of CO_2 (mainly from fossil fuels and the burning of forests) will have to be cut as much as 50 to 80 percent to avoid major climate change.

Such a reduction in the CO_2 emission rate probably cannot be accomplished without a massive switch to nonfossil-energy sources (see Figure 8.4). Nevertheless, emissions from fossil fuels can be moderated somewhat by three strategies: exploiting the fuels more efficiently, replacing coal by natural gas and recovering and sequestering CO_2 emissions. Most important is efficiency improvement, because it is often economically and environmentally attractive and because the opportunities are plentiful for all uses and for all nations.

The natural gas option complements improvements in efficiency, and it, too, is appealing for various reasons. Compared with coal, combustion of natural gas, which is mainly methane, yields about 70 percent more energy for each unit of CO_2 produced. Also, natural gas can be burned efficiently because of the simplicity of gas-handling equipment, because it lacks ash (unburnable material) and because it typically contains much less sulfur than is found in coal (hence, energy does not have to be diverted to cleanup devices). Moreover, advanced power-generating methods, some of which derive from jet engine technology, are now being developed that are well suited to natural gas.

In one such system, known as STIG (for steam-injected gas turbine), natural gas (or synthesis gas from coal) is burned to drive a gas turbine to pro-

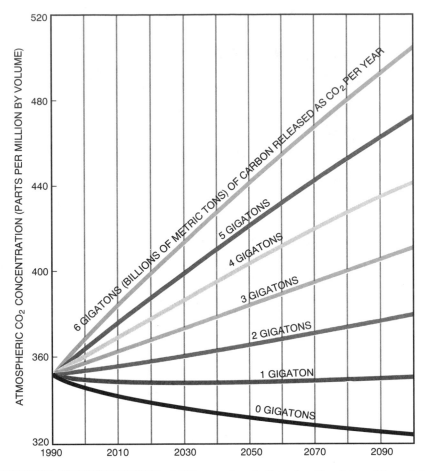

Figure 8.4 ATMOSPHERIC CONCENTRATION of CO₂, which contributes to global warming, has been projected by William R. Emanuel of the Oak Ridge National Laboratory. He assumed that the current rate of CO_2 emissions from human activities would suddenly change to the values indicated on the plots and be maintained thereafter. Today's emission rate of close to six gigatons of carbon per year must drop to about one gigaton per year, roughly a sixth of what it is now, to stabilize the atmospheric concentration, a change that is not easily accomplished.

duce electricity. The exhaust gases from the turbine are then channeled to a boiler to produce steam not for a second turbine (as in combined cycles) but for injection into the combustion chamber. The steam injection increases the energy output and efficiency while also reducing NO_x emissions.

Further efficiency can be gained if the combustion air, which has to be compressed, is cooled between stages of compression, with some of the air channeled to the turbine blades. Efficiency is gained because less power is required to compress cool air. Furthermore, the diverted air lowers the temperature of the blades, enabling them to withstand higher gas temperatures. Because such intercooled

STIG (ISTIG) systems can be small, they are suitable for industry as well as utilities (see Figure 8.5).

A proposed variation on the ISTIG theme might yield an efficiency of more than 52 percent. The approach involves the precombustion catalytic reaction of natural gas with steam to yield H₂ and CO. The chemical energy of the products would be greater than that of the natural gas itself.

Although the technologies for using natural gas efficiently are improving (see Figure 8.6), there are some significant limitations to the strategy of substituting gas for coal. First, natural gas is much less abundant. At the current worldwide rate

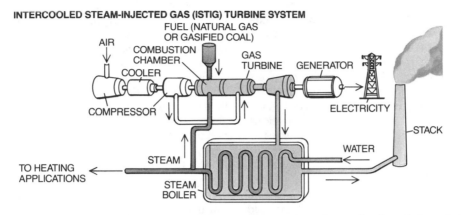

INTERCOOLED STEAM-INJECTED GAS (ISTIG) TURBINE SYSTEM

Figure 8.5 ISTIG SYSTEM is well suited to natural gas. Like certain systems that run combined cycles, it has a gas turbine and captures heat from the turbine to produce steam. The steam, however, is channeled to the combustion chamber (increasing power output and efficiency). At the same time, the combustion air is cooled during compression, making that process easier and reducing the power needed to run the system. Diversion of air from the compressor unit to the turbine blades cools them, thus enabling them to withstand higher gas temperatures.

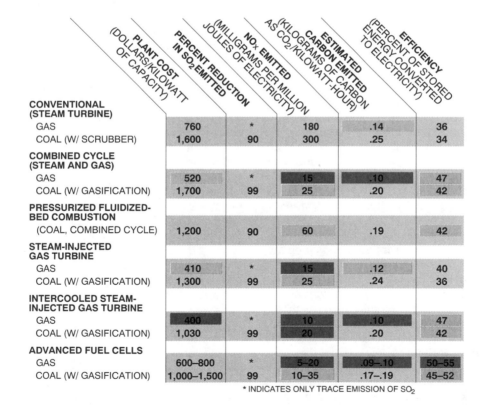

	PLANT COST (DOLLARS/KILOWATT OF CAPACITY)	PERCENT REDUCTION IN SO₂ EMITTED	NOx EMITTED (MILLIGRAMS PER MILLION JOULES OF ELECTRICITY)	ESTIMATED CARBON EMITTED (KILOGRAMS OF CARBON AS CO₂/KILOWATT-HOUR)	EFFICIENCY (PERCENT OF STORED ENERGY CONVERTED TO ELECTRICITY)
CONVENTIONAL (STEAM TURBINE)					
GAS	760	*	180	.14	36
COAL (W/ SCRUBBER)	1,600	90	300	.25	34
COMBINED CYCLE (STEAM AND GAS)					
GAS	520	*	15	.10	47
COAL (W/ GASIFICATION)	1,700	99	25	.20	42
PRESSURIZED FLUIDIZED-BED COMBUSTION					
(COAL, COMBINED CYCLE)	1,200	90	60	.19	42
STEAM-INJECTED GAS TURBINE					
GAS	410	*	15	.12	40
COAL (W/ GASIFICATION)	1,300	99	25	.24	36
INTERCOOLED STEAM-INJECTED GAS TURBINE					
GAS	400	*	10	.10	47
COAL (W/ GASIFICATION)	1,030	99	20	.20	42
ADVANCED FUEL CELLS					
GAS	600–800	*	5–20	.09–.10	50–55
COAL (W/ GASIFICATION)	1,000–1,500	99	10–35	.17–.19	45–52

* INDICATES ONLY TRACE EMISSION OF SO₂

Figure 8.6 COMPARISON of advanced fossil-fuel technologies reveals that, as a rule, gas-fueled systems are less expensive to install, emit less sulfur dioxide (SO₂) and CO₂ and are more efficient than their coal-driven counterparts (although fuel costs can be higher). Red, blue and green colors highlight, in order, the first, second and third best numbers in a column. All values, except for the conventional steam plant, are optimistic, but achievable, numbers in the literature; those for the conventional system reflect average values for existing plants. Efficiencies were calculated on a higher heating value basis, the U.S. convention.

of consumption, and with existing technology, estimated economically recoverable resources of coal would last perhaps 1,500 years. In contrast, reserves of natural gas would last only 120 years (or perhaps two or three times that long if the much higher costs of recovering less accessible gas from unconventional sources were accepted or if methods were developed to recover it more economically). If natural gas was substituted for coal in all applications, the gas resources might last only 55 years (see Figure 8.7).

Second, leakage of natural gas during extraction and transport could partially offset the advantage of its use, because CH_4 is a greenhouse gas. In fact, methane is a more efficient absorber of infrared radiation than is CO_2, although its residence time in the atmosphere is much shorter.

The leakage problem should have a technical solution, but what can be done about the supply problem? That concern diminishes somewhat if one recalls that gas would be needed only until competitive nonfossil technologies became established.

The uneven distribution of gas resources makes matters more problematic. For instance, U.S. supplies would last only about 18 years if natural gas replaced coal in all uses, but those of the Soviet Union would last 70 to 80 years. Such uneven distribution raises concern about access. At least on a technical level, that problem seems solvable, however. Although most natural gas is used in the countries where it is produced, trade has been increasing since the early 1970's and is expected to continue to do so, in part because the World Bank and other agencies are supporting exploration for natural gas and the building of distribution infrastructures in the developing nations.

Common wisdom holds that much more gas will be found as exploration for it becomes as profitable as it is for oil. Today world trade via pipeline accounts for about 11 percent of total use, and tanker shipments of liquefied natural gas (LNG) account for about 3 percent. The potential for growth of a global trade network is evident from even a cursory glance at the pieces already being fit into place. The most extensive distribution network in the world delivers natural gas to the U.S., which is linked to supplies from Canada by pipeline and from North Africa by tankers carrying LNG. There is also a pipeline from Mexico that could be put into operation if market conditions justified such a move. Plans are being considered that would bring LNG to the U.S. from elsewhere.

A network in Western and Eastern Europe is also well developed and receives supplies from the Netherlands, Norway, North Africa and the U.S.S.R. It might be improved further now that Iran is planning to add another pipeline to ones already serving the Soviet Union. In exchange for the Iranian gas, the U.S.S.R. will supply its own gas to Western and Eastern Europe. The Soviet Union has also recently initiated discussions with Japan about building a 3,100-mile (5,000-kilometer) pipeline, partly undersea, between Siberia and Japan via South and North Korea. Eventually the Soviet Union could supply gas to China as well.

Japan, too, has made considerable progress toward ensuring itself a supply of natural gas; it imports the liquefied form from Indonesia, Malaysia, Australia, Alaska and the Middle East. In addition, Iran is promoting a $12-billion project to bring gas to India and Pakistan by a 2,000-mile pipeline from Bandar Abbas to Calcutta.

Whether the nations of the world are willing to make a major transition from coal to natural gas is another question. At the moment, gas is more expensive to buy than coal and may become more so as demand increases. Also, the U.S. and other countries that have abundant supplies of coal and smaller supplies of gas might be unwilling to increase significantly their dependence on outside sources of the fuel. Every nation is, after all, concerned about having secure and stably priced sources of energy.

On the other hand, the expanded use of natural gas is quite viable for some regions, such as Eastern and Western Europe. If the Soviet Union became a major supplier of gas to those areas, the Soviet economy would be greatly strengthened, and both CO_2 emissions and pollution would be reduced. Helping the U.S.S.R. with advanced gas production, transmission and utilization technologies would seem to be a stabilizing policy for the industrialized countries of the West to adopt in this day of *perestroika*.

Similarly, encouraging the Middle East to expand its exports of natural gas not only to Europe but also to Pakistan and India would seem a wise environmental policy. And helping the Pacific Rim nations to use the gas resources of such countries as Indonesia, Malaysia and Thailand would be an environmentally and economically desirable development strategy.

A third, more futuristic, strategy for reducing CO_2 emissions from fossil fuels is to capture the emis-

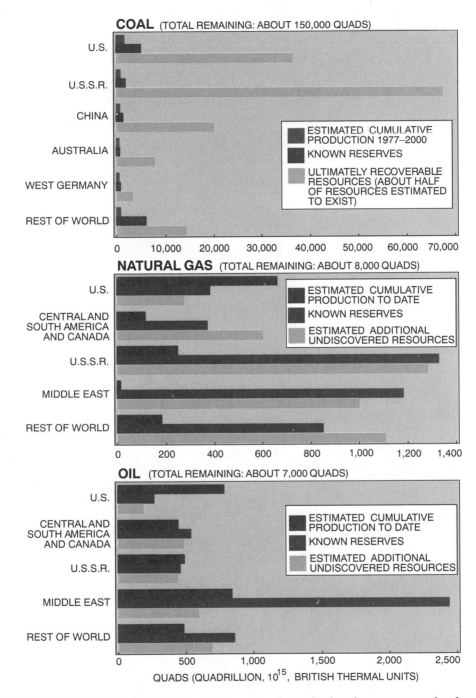

Figure 8.7 ECONOMICALLY RECOVERABLE fossil fuels are not distributed evenly. At 1988 use rates, stores of coal, natural gas and oil could last 1,500, 120 and 60 years, although production from unconventional sources could double the time for gas and oil.

sions and sequester them. One might, for instance, plant forests to absorb an amount of the gas equivalent to what is emitted (see Figure 8.8). Yet offsetting the emissions of a single, 500-megawatt coal-fired power plant (a typical size for a utility) operating at 34 percent efficiency would require planting a forest of about 1,000 square miles. Sequestering all the carbon emitted as CO_2 by fossil fuels in the U.S. in 1988 — 1.4 gigatons (billion metric tons) — would require a million square miles of new forests, roughly 25 percent of the land area of the U.S. Moreover, an afforestation plan would have to include means of preserving new forests. Nevertheless, the costs of afforestation are lower than those for any other method of CO_2 sequestration under consideration (see Figure 8.9).

Alternatively, CO_2 can be recovered directly from the emissions of large point sources, such as power plants, and stored. Recovery can be done by various means, such as chemical absorption. Then the gas would be compressed for transport to a storage site. Storage options include piping the gas deep into the ocean or into depleted natural gas reservoirs.

The whole proposition would probably be extremely expensive, at least for the methods people have dreamed up so far. Part of the problem is the enormity of the emissions: burning one ton of carbon in fossil fuel gives rise to more than three and a half tons of CO_2.

Furthermore, no one knows the effects of ocean storage on the ecology of the sea or how long the gas would remain there or in gas reservoirs. Also, the storage capacity in depleted gas reservoirs would probably not become available on a large

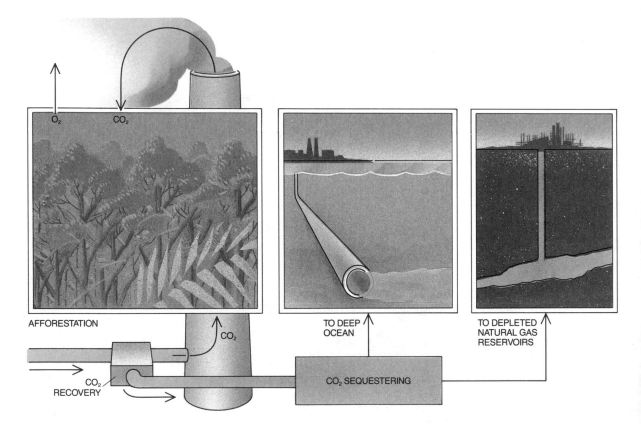

Figure 8.8 SCHEMES for controlling carbon dioxide emissions include afforestation—the planting of forests to absorb CO_2 from the atmosphere (*left*). Alternatively, the gas might be captured before it reaches the air and then be piped deep into the ocean (*center*) or possibly into natural geologic formations such as reservoirs of natural gas (*right*).

REMOVAL PROCESS	PERCENT OF CO_2 REMOVED	COST (CENTS PER POUND OF CARBON AS CO_2)		PERCENT REDUCTION IN EFFICIENCY	COST OF ELECTRICITY RELATIVE TO TODAY'S COST
		REMOVAL	SEQUESTERING		
MOLECULAR SIEVES	90	8.4	1.1 *	80	5.0
LIME ADSORPTION	90	5.9	1.1 *	59	2.4
MEA SCRUBBING	90	5.5	1.1 *	57	2.3
OXYGEN COMBUSTION	100	7.0	1.1 *	30	1.8
IGCC/ SELEXOL	88	1.8	.7 **	13	1.3
AFFORESTATION	100	0	.9	0	1.1

* —IN DEEP OCEAN   ** — IN DEPLETED NATURAL GAS WELL

Figure 8.9 ESTIMATED COSTS of proposed methods of recovering and sequestering CO_2 from large coal-fired electric-power plants vary enormously. CO_2 is recovered from gasified coal (in an IGCC system) by a sorbent called selexol. Molecular sieves are porous materials permeable to some gases but not others; MEA is the chemical monoethanolamine. These materials can physically or chemically capture the CO_2 from the exhausts of a conventional plant. Oxygen combustion means oxygen replaces air in the boiler; the process eases CO_2 recovery because it results in a high concentration of the gas in the exhausts. Afforestation may be the least expensive option but, because of the huge land requirements, is probably impractical on a large scale.

scale for 20 to 30 years and would then be filled relatively quickly, in another 20 to 30 years. And the approach would be practical only for large, stationary sources of CO_2, which in the U.S. account for only about 30 to 40 percent of emissions.

Nevertheless, the notion is worth considering because of the implications for fossil-fuel use and the environment. Although various early analyses suggested that the capture and storage of CO_2 would about double the cost of electricity, a recent calculation by Chris A. Hendriks, Kornelis Blok and Wim C. Turkenberg of the University of Utrecht in the Netherlands lowers the figure considerably. The team suggests the cost of electricity could rise only 30 percent, which puts it in the realm of feasibility.

The Utrecht group assumed an oxygen-blown gasifier would be operated to maximize H_2 production and that the H_2 would serve as fuel in a combined cycle. The CO_2 formed during the H_2 production process would be compressed and sequestered in a depleted natural gas reservoir 100 kilometers away. Some observers suggest that 50 to 100 years from now hydrogen, perhaps generated from water with power provided by the sun or by nuclear fission or fusion, will be a major nonpolluting energy carrier. With that possibility in mind, Robert H. Williams of Princeton University has suggested that production of H_2 from gasified coal might be a nontraumatic way to begin a transition to the widespread use of hydrogen as an energy source. The H_2 could be produced near coal fields and piped to centers of fuel use, where it might power very efficient gas turbines or fuel cells. At the same time, the CO_2 produced with the H_2 would be piped to sequestering locations.

Fuel cells operate something like a battery, except that the fuel is constantly replenished. In the phosphoric acid cell, the only one in commercial use, the H_2 first encounters a permeable anode and is oxidized (electrons are removed), yielding positively charged hydrogen ions (H^+) and free electrons. Then the electrons flow in an external circuit, providing power. Meanwhile the hydrogen ions migrate through phosphoric acid. The electrons and ions meet at the cathode and react with oxygen to form steam, which can be harnessed for heating or, less efficiently, discarded.

To produce the electricity needed by a utility, stacks of fuel cells would constitute the power-generating unit. At least three types of cell aside from the phosphoric acid variety are being studied closely. These devices may operate at efficiencies as high as 50 to 60 percent, assuming the heat produced is exploited.

It is clear that many good minds are focused on the dual problems of improving energy generation from fossil fuels and replacing the fossil-fuel systems of the world with environmentally friendlier options. Yet many problems remain to be solved. For instance, the world's energy systems have great inertia and thus tend to change quite slowly, because it is costly to abandon expensive equipment before its useful life is completed. Hence, even if the world became convinced that a reduction in greenhouse emissions was necessary, the reduction could take decades to accomplish.

This problem of inertia is related to another one. If a greenhouse effect actually imposed a sudden constraint on CO_2 emissions, the costs of responding immediately would be extremely high not only because of unreturned investments on the original equipment but also because the nonfossil technologies developed to date are still expensive and limited in supply. Furthermore, the costs would be unevenly distributed and would be most oppressive for the developing nations, because they are least able to pay and because their energy needs are growing rapidly.

Some computer models suggest that a global release "ration" of 1.6 to 3.5 gigatons of carbon as carbon dioxide per year would result in a safe rate of warming—on the order of .05 degree Celsius a decade—assuming other greenhouse gases were also controlled so that they contributed only another .05 degree C. If that assessment is correct, the question becomes who gets to use what? And does the ration go preferentially to particular uses?

Industrialized nations, which make up only about a third of the world population, generate about 80 percent of the greenhouse gases. The developing countries, on the other hand, consume much less energy per person. Do the industrialized nations reduce their consumption rates so that the developing nations can increase theirs and thus have further economic growth?

The needs of the developing nations bring up a final issue. A simple extrapolation of the rates of energy use in the past decade leads to the conclusion that the CO_2 emissions in the developing countries (including China) could exceed those of the developed nations during the next decade. Some say the installation of efficient technologies could allow the developing countries to achieve a level of affluence equivalent to that of Western Europe in the mid-1970's but with a rate of energy consumption per person that would be about one third that in Western Europe today. For this to happen, the developing nations must be willing to invest money to save more later. Yet the difficulties of raising capital encourages the purchase of less expensive, but less efficient, equipment.

Many of these issues have no clear solutions, but some recommendations for the immediate future can be made nonetheless. It seems prudent to manage fossil fuels as if a dangerous greenhouse warming were probable. Such an undertaking involves adopting more efficient technologies across the board—which makes good economic sense.

It also makes sense to substitute natural gas for coal wherever doing so is practical. Although the consumption of natural gas instead of coal might increase natural gas prices and hasten depletion of reserves, it may also buy time, affording the environment some protection until technologies for nonfossil-energy sources are perfected. On the other hand, oil prices fell throughout the 1950's and 1960's despite a growth in demand. Internationally traded supplies of natural gas might well increase markedly without a rise in prices if the abundant resources of the Soviet Union and the Middle East are tapped. In fact, we think the Western industrialized nations should cooperate in providing technical assistance and capital to developing countries and Eastern Europe as a way of encouraging the development of natural gas resources and the adoption of more efficient energy technologies.

Concurrently, there should be more research into the global carbon cycle, so that well-informed targets for greenhouse emissions can be established. And there must be more research into efficient fossil-fuel technologies (such as gas turbines and fuel cells), into methods for producing H_2 from coal and for recovering and sequestering CO_2 (at low cost and with minimal environmental consequences), as well as into highly effective, safe and economical ways to harness energy from nonfossil sources.

Until such research makes more progress, it will probably remain hard to wean the world from fossil fuels. For all their faults, they remain relatively inexpensive, widely available and readily adaptable to applications large and small, simple and complex.

Energy from Nuclear Power

Atomic energy's vast potential can be harnessed only if issues of safety, waste and nuclear-weapon proliferation are addressed by a globally administered institution.

. . .

Wolf Häfele

Nuclear power has been hailed as the solution to the world's energy problems and condemned as the most dangerous and unfitting way to produce energy. Today, in the aftermath of the Chernobyl and Three Mile Island accidents, public opposition has strengthened. Reactor construction has come to a halt in many countries: in the U.S. there have been no new orders since 1978; in Sweden a referendum has called for the termination of nuclear-power generation by 2010; in Switzerland and West Germany there is a de facto moratorium (see Figure 9.1), and in the Soviet Union debate about the future of nuclear power is intensifying.

The key to the sometimes bitter controversies surrounding nuclear power lies in its two opposing physical characteristics. On the one hand, nuclear energy can provide a trillion times more energy than mass forces such as wind and water and a million times more energy than the chemical reactions of the industrial revolution (chiefly combustion and electrochemistry), which led to the transformation of society. Current levels of nuclear-power production do not live up to this potential—which I refer to as the one trillion factor.

On the other hand, chemical and nuclear power generate waste. As John P. Holdren of the University of California at Berkeley has observed, mass forces do not fundamentally change the molecules they act on, and so harnessing wind and water is inherently environmentally benign. In contrast, chemical and atomic reactions change molecules and nuclei, respectively, thereby creating waste. Advocates for the use of mass forces must identify ways to overcome the limitations of the proportionately small amount of energy produced by those means. Those encouraging greater reliance on nuclear power and chemical energy must address the problems of waste—both radioactive and atmospheric.

Clearly, nuclear energy must be critically assessed. Many future scenarios show that meeting energy needs and environmental demands without nuclear power will be difficult. At the same time, an ever greater role for nuclear energy appears inconceivable to some given the current climate of political opposition.

These political and social impasses could be overcome if global institutions and provisions are established. Any international organization should reflect a 1946 proposal made by then U.S. Secretary of State Dean G. Acheson and David E. Lilienthal

and Bernard M. Baruch of the United Nations Atomic Energy Commission, which called for the creation of a global agency to govern atomic power. Such an agency, tailored to the concerns of the year 2000 and beyond, could ensure reactor safety and implement and monitor waste storage facilities. Nuclear power must be considered as part of an overall energy system, one that renders human activity compatible with a sustainable environment.

Before we address the future, we must examine the current situation of nuclear power. By 1989 nuclear energy accounted for 16 percent of the world's electricity. By the end of the year 426 nuclear reactors were connected to the global electric grid (see Figure 9.2), according to the International Atomic Energy Agency (IAEA), with a net generating capacity of 318,271 megawatts of electricity (MWe). (Equivalent to one gigawatt, 1,000 MWe can supply enough electricity for some 500,000 homes.) Ninety-six reactors are under construction, which when completed will bring the total nuclear generating capacity to 397,178 MWe—roughly 400 gigawatts electric (GWe).

Twelve reactors (net capacity 10,407 MWe) were added to the global grid in 1989, and 15 such connections are expected in 1990 (net capacity 12,151 MWe). Although many Europeans and Americans believe nuclear power to be at a standstill, construction was started on five reactors in 1989 (net capacity 4,738 MWe): two in Japan, two in South Korea and one in the Soviet Union.

Let us consider what would happen if the level of electricity generated by nuclear fission were held constant at 400 GWe for 100 years—if the nuclear plants now in operation or under construction represent the final world total. Some 30 tons of 3 percent enriched uranium 235 are needed to produce one gigawatt over the course of a year in a contemporary light-water reactor (LWR)—so called because ordinary water is used as a coolant and is also converted to steam by the heat generated during fission. Seventy-five percent of the reactors operating today are LWR's. (Uranium 235, which occurs in natural uranium 238, is concentrated, or enriched, so

that critical mass—the initiation of the fission chain reaction—can be achieved in the reactor.)

In the likely event that a 15 percent improvement is attained in core performance—meaning that the fuel elements will have a longer lifetime—only 20 tons annually (equal to 160 tons of unenriched, that is, natural uranium) would be needed. So, if conservative assumptions are correct, between six and seven million tons of uranium—equivalent to the total currently estimated world supply—will be required to meet generating capacity over the next century.

This model, however, is based on a once-through, throwaway cycle in which fuel is not reprocessed to be used again, and so it results in a buildup of irradiated fuel elements. Because the amount of fresh fuel used in a reactor equals the amount of spent fuel that is discharged, 400 GWe would produce 20 tons of irradiated fuel elements (94 percent of which is uranium; 1 percent its by-product, plutonium; and 5 percent fission products)—an annual total of 8,000 tons (see Figure 9.3). And because waste disposal sites would store an estimated 70,000 tons of heavy metal (principally uranium and plutonium) per site, every ninth year such a site would have to be opened (11 such sites over 100 years).

Disposal repositories are technically uncomplicated, but given public opposition to the handling and disposal of radioactive materials, they pose significant legal and societal problems. Already just a few such sites, including Yucca Mountain in the U.S. and Gorleben in West Germany, have been delayed and canceled because of strenuous protest. To date, no long-term sites exist anywhere in the world.

In addition to consuming the world's supply of uranium and producing a mountain of irradiated fuel elements over the next century, LWR technology would produce comparatively little energy. Despite the enormous amount of energy obtainable through nuclear power as opposed to other sources, a scenario of 400 GWe is equivalent to only 70 terawatt-years of thermal energy (TWth), far less than the already discovered reserves of oil (154 TWth) or gas (130 TWth). The reason is that even if the efficiency of LWR's is improved 15 percent, they will fission only .6 percent of the available uranium atoms; 99.4 percent are wasted.

Figure 9.1 NUCLEAR REACTOR in Leibstadt, Switzerland, commenced commercial operation in 1984. The boiling-water reactor (BWR) is a form of light-water reactor (LWR), which drives turbines with heat from fission captured by steam.

Even so, the greenhouse effect offers a powerful argument for increasing the use of nuclear

	REACTORS IN OPERATION	ELECTRICITY GENERATED (MWe)	PERCENT OF ELECTRICITY GENERATION	REACTORS UNDER CONSTRUCTION
NORTH & CENTRAL AMERICA				
Canada	18	12,185	15.6	4
Cuba	0	0	0	2
Mexico	1	654	—	1
U.S.	110	98,331	19.1	4
SOUTH AMERICA				
Argentina	2	935	11.4	1
Brazil	1	626	.7	1
EUROPE				
Belgium	7	5,500	60.8	0
Bulgaria	5	2,585	32.9	2
Czechoslovakia	8	3,264	27.6	8
East Germany	6	2,102	10.9	5
Finland	4	2,310	35.4	0
France	55	52,588	74.6	9
Hungary	4	1,645	49.8	0
Italy	2	1,120	—	0
Netherlands	2	508	5.4	0
Romania	0	0	0	5
Spain	10	7,544	38.4	0
Sweden	12	9,817	45.1	0
Switzerland	5	2,952	41.6	0
U.K.	39	11,242	21.7	1
West Germany	24	22,716	34.3	1
Yugoslavia	1	632	5.9	0
ASIA				
China	0	0	0	3
India	7	1,374	1.6	7
Iran	0	0	0	2
Japan	39	29,300	27.8	12
Pakistan	1	125	.2	0
South Korea	9	7,220	50.2	2
Taiwan	6	4,924	35.2	0
U.S.S.R.	46	34,230	12.3	26
AFRICA				
South Africa	2	1,842	7.4	0
TOTALS	426	318,271	—	96

Source: International Atomic Energy Agency, Vienna

Figure 9.2 AMOUNT OF POWER produced by reactors worldwide is shown in megawatts electric, along with the percentage of electricity each one provides. In the last column the reactors under construction as of December 31, 1989, are given.

power—if the safety and waste problems can be overcome. Various carbon dioxide (CO_2) reduction scenarios have been proposed to offset the possible consequences of global warming, and each one presents a different role for nuclear power. The 1988 Toronto conference, "The Changing Atmosphere: Implications for Global Security," called for CO_2 reductions of 20 percent, from six billion tons of carbon to 4.8 billion tons a year by 2005. At the 14th World Energy Conference in Montreal in 1989, I proposed a further reduction of energy-related CO_2 emissions—to four billion tons a year by 2030. My so-called Jülich CO_2 reduction scenario called for increased energy conservation, a shift from coal to natural gas and greater dependence on solar power and biomass combustion. Nuclear power provided the remainder of the energy.

Unlike other scenarios (see Figure 9.4), the Jülich

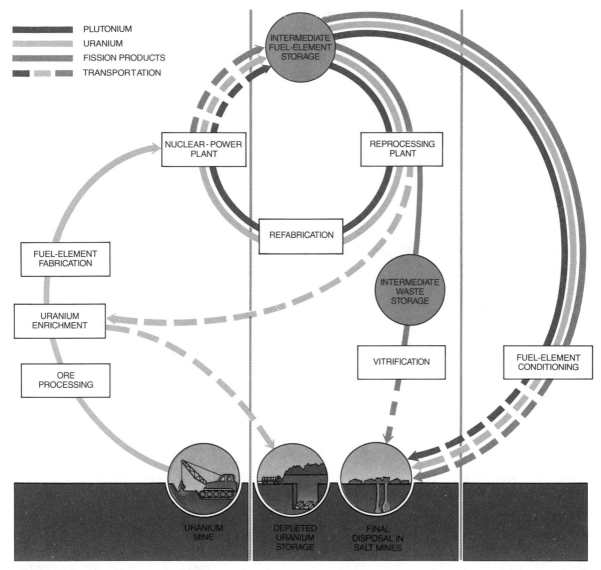

PLUTONIUM
URANIUM
FISSION PRODUCTS
TRANSPORTATION

INTERMEDIATE
FUEL-ELEMENT
STORAGE

NUCLEAR-POWER
PLANT

REPROCESSING
PLANT

REFABRICATION

FUEL-ELEMENT
FABRICATION

INTERMEDIATE
WASTE
STORAGE

URANIUM
ENRICHMENT

ORE
PROCESSING

VITRIFICATION

FUEL-ELEMENT
CONDITIONING

URANIUM
MINE

DEPLETED
URANIUM
STORAGE

FINAL
DISPOSAL IN
SALT MINES

Figure 9.3 NUCLEAR FUEL CYCLE follows the life history of uranium, the principal element used as fuel in nuclear fission. Uranium occurs naturally but in most cases needs to be enriched (a process in which the small percentage of fissile uranium is concentrated) before it can be used as fuel. In a once-through cycle, spent fuel elements can be disposed of in retrievable or nonretrievable storage facilities. Otherwise, spent fuel can be chemically reprocessed and, after fabrication, reused. Plutonium generated during fission is disposed of or recycled after reprocessing for reuse in fuel.

proposal considers the rising cost of energy conservation. Conservation is not a controversial solution for energy shortages—to a point. But high energy savings require a disproportionately large capital investment as well as changes in the existing infra-structure. These investments ultimately make the financial and societal costs of generating energy cheaper than the costs associated with saving it. With these limitations in mind, I find the proposal by Umberto P. Colombo (of the Italian Commission

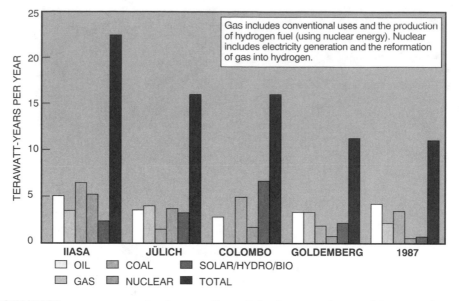

Figure 9.4 SCENARIOS compare amounts of energy (in terawatt-years per year) provided by different sources in 1987, with proposals for 2030. The 1975 International Institute for Applied Systems Analysis and the 1979 Umberto P. Colombo scenarios predate current greenhouse-effect models; the 1989 Jülich scenario and the 1988 José Goldemberg proposal (for 2020) take climate change into account.

for Atomic and Alternative Energy Sources) for 43 percent energy savings for all industrialized countries to be the feasible limit of conservation; Secretary of Science and Technology for Brazil José Goldemberg's estimate of 69 percent is too extreme.

Given the financial and societal constraints imposed on energy conservation, I propose that nuclear power increase from its present level of 400 GWe to two terawatts (1 TWe of electricity equals 1,000 GWe) by the year 2030 (more thereafter), accompanied by an evolution in current reactor types as well as the development of more efficient technologies. The greater energy capacity (equivalent to 3.7 TWth) would be used both for the generation of electricity as well as for the reformation of natural gas into hydrogen—a potentially important fuel (see Chapter 8, "Energy from Fossil Fuels," by William Fulkerson, Roddie R. Judkins and Manoj K. Sanghvi). Such a scenario is not inconsistent with calls for a revisiting of nuclear power as requested in the Toronto conference's final statement.

Even without considering the threat of global warming alone, augmenting levels of nuclear-power generation makes sense. My proposal reflects 1975 and 1986 estimates by the International Insti-

tute for Applied Systems Analysis and the Organization of Economic Cooperation and Development (OECD). Economic and overall environmental prudence together argue for increased nuclear dependence, if for no other reason than to handle the vicissitudes of world energy supplies.

Generating two terawatts calls for a total of 2,500 nuclear reactors as well as assurances of safety, adequate waste disposal, stringent safeguards against the proliferation of nuclear weapons and, above all, public acceptance. Although such a large buildup sometimes appears impossible, past construction records suggest otherwise. Roughly 40 GWe would have to be added to the global electric grid every year to reach a generating capacity of two terawatts. In 1984 and 1985 more than 30 GWe were added; 40 GWe are therefore clearly within reach.

From its beginnings in the 1950's the nuclear industry has meticulously examined safety features, but the prevailing concept of safety analysis has changed. Today's predominant approach to safety is known as probabilistic risk analysis (PRA). PRA uses as a definition of risk the probability that an event will happen multiplied by its consequences. The definition was borrowed from the insurance industry, which treats risk as the probability of an event multiplied by its monetary cost.

PRA's application to nuclear safety, however, is not without shortcomings. Indeed, PRA's failure to describe fully all the implications of large nuclear accidents is partly responsible for the lack of public confidence in nuclear power. Further, the probability of a catastrophic event is intended to be an improbability, and statistics are not available in the absence of such events. Instead a multiplicity of statistics synthesizes a big event from many component failures—and in so doing creates additional uncertainty.

Additionally, these improbabilities imply long periods, if not eternities, until the occurrence of a catastrophe. But the logic of probability does not incorporate the notion of eternity, and so the damages of major catastrophes escape rational assessment. PRA focuses on catastrophic events multiplied by low probabilities, when in reality only the consequences—not the probability—of such events are understandably considered by the public.

To be meaningful, PRA's need to be comparative. For example, comparing nuclear reactor designs has successfully led to the identification and correction of design flaws. Such analyses are remarkably sensitive and sophisticated in terms of identifying possible safety lapses. By applying PRA to the design of modern nuclear reactors, engineers estimate that a major LWR core meltdown is likely to happen only once in 20,000 years. If a meltdown takes place, the probability of a related release of radioactivity can be reduced still further by a well-designed containment facility. Thus, a major leak of radioactivity would be expected only once every, say, 100,000 years.

When PRA is applied to the 400-GWe scenario, one massive release of radioactivity is expected to occur every 200 years. In the two-terawatts scenario, such a release would occur every 40 years because there would be five times as many reactors and therefore five times as many chances for mishaps. Clearly, such a frequency is not acceptable, so it is imperative that the safety of reactors outpace their numbers. I envisage the probability of a core meltdown held within a containment to be less frequent than, say, once every 4,000 years.

But even when this infrequency is compared with the probability of accidents associated with other energy sources, an explosion at an oil refinery, for example, the fear of a nuclear mishap persists—especially after the accidents at Chernobyl and Three Mile Island, near Harrisburg, Pa. But at Harrisburg the containment structure worked—there was no appreciable leak of radioactivity on March 28, 1979—whereas at Chernobyl, on April 26, 1986, there was almost no containment and massive contamination.

So we must strive for the construction of containments that prevent the damages associated with meltdown or other nuclear accidents. To achieve effective containment, I suggest we not look at the risk of the event but consider frequency and consequence separately. Therefore, we must limit, through design, the effects of a severe accident no matter how low its probability. A properly designed and constructed containment would protect people and the environment outside the reactor by limiting the outcome.

Stringent containment is one element necessary to all large-scale applications of nuclear power. New designs permit the "deterministic" exclusion of catastrophic consequences such as a core melt-through (when the fuel overheats and melts through the floor of the reactor) or steam explosion. Hans Henning Hennies, Günter Kessler and Josef Ebil of the University of Nuclear Research Center of Karlsruhe, West Germany, for example, recently analyzed the feasibility of an improved LWR containment design (see Figure 9.5). At first glance, existing containments have much in common with this design, but the key difference is that the Karlsruhe design goes farther with each feature.

The design provides for emergency cooling through the natural circulation of coolant between the concrete shell and its steel lining—even under high pressure. Any emissions that might enter and escape the coolant are filtered at the top of the dome. A core catcher is provided in the event of a melt-through; long steel cables would absorb and dissipate any energy from an exploding core fragment.

Another safety improvement is the introduction of so-called passive features. In passively safe reactors, several days should be able to pass before human intervention is required to contain radioactivity. Features include a large capacity to absorb heat from the reactor's core and the network of pipes that pumps coolant to the reactor. It should also have a large ratio of cooling surface to core volume that allows the core to cool by heat diffusion and natural convection. Other designs strive

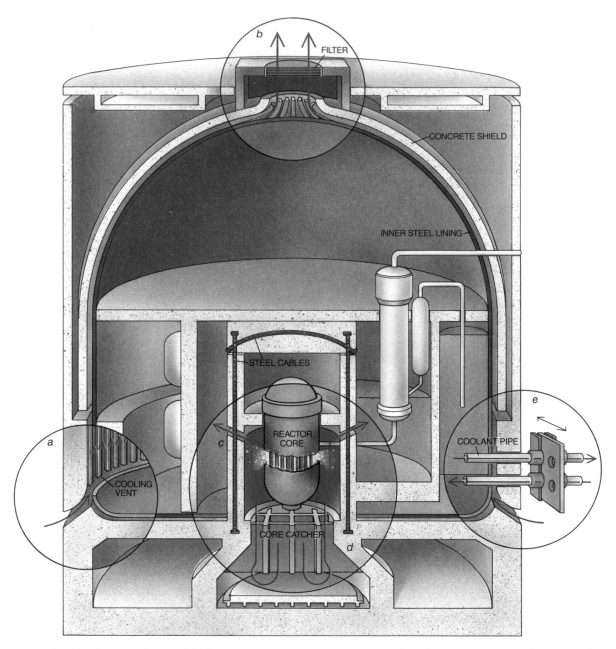

Figure 9.5 CONTAINMENT DESIGN for an LWR ensures that core fragments and radioactivity stay within the power plant in the event of an accident. The concrete structure has an inner steel lining 40 millimeters (about one and a half inches) thick that provides cooling vents for natural air circulation (*a*). Air flows to the top of the dome, where a filter prevents the escape of radiation (*b*). If the reactor core explodes, long steel cables absorb and dissipate the energy from the upward-moving fragments (*c*); a grid and core catcher below the core prevent melt-through to the ground (*d*). In the event of pressure buildup, sliding gates close off all pipes, preventing the release of radioactivity (*e*). The drawing is based on a design from the University and Nuclear Research Center of Karlsruhe, West Germany.

for simplicity, thereby improving safety in other ways as well [see "Advanced Light-Water Reactors," by Michael W. Golay and Neil E. Todreas; SCIENTIFIC AMERICAN, April, 1990].

In addition to safety, a two-terawatts scenario must take into account fuel supply. Changing to an efficient fuel cycle where 60 percent instead of .6 percent of the uranium is fissioned becomes imperative. As I mentioned earlier, the throwaway fuel cycle of 400 GWe would exhaust global uranium supplies in 100 years. At a generating capacity of two terawatts, uranium would be depleted in 20 years.

The fuel-efficiency technology available today is the fast breeder reactor (FBR)—so called because it creates more fuel than it consumes by converting nonfissile uranium into fissile plutonium. FBR's have passed the test of engineering feasibility and are on their way toward commercial feasibility. Because they can extract 100 times more energy from the same amount of uranium as LWR's, they represent an important technological breakthrough. If all uranium were fissioned in FBR's, nuclear power could generate at least 35,000 TWth as opposed to the 70 TWth now being generated (partly because the higher cost of additional low-grade uranium, which is much harder to mine, could be balanced by the savings from breeder efficiency).

Breeders require reprocessing. If FBR's and LWR's generate two terawatts, they would produce 35,000 tons of irradiated fuel elements. In this scenario a final waste disposal site would have to be opened every other year or so—an awkward prospect. In the reprocessing process, plutonium created during the fission of uranium is extracted chemically from the spent fuel along with the unused uranium. (Spent fuel is dissolved in nitric acid, which forms a solution from which uranium and plutonium and the fission products—nuclear ash—can be removed, also by chemical means.)

The four reprocessing facilities currently operating—Sellafield in England, La Hague in France, Tarapur in India (see Figure 9.6) and Tokai in Japan—process a total of 2,040 tons of uranium a year. Four plants under construction will process another 2,500 tons a year, yielding a global reprocessing capacity of 4,540 tons a year. (There are no data available on Soviet or Chinese reprocessing capabilities.) These existing and planned facilities can only reprocess half of the 10,000 tons of spent fuel currently produced each year—nowhere near the above-mentioned 35,000 tons a year.

Beyond safety features and efficient fuel cycles wait other technological developments, including fusion. Fusion releases energy by uniting nuclei rather than dividing them. Although fusion has not yet passed the threshold of scientific feasibility and cannot be incorporated into future scenarios, it holds great promise. The intended International Tokamak Reactor Experiment (a joint enterprise involving the U.S., the Soviet Union, Japan and Europe) suggests that by the middle of the next century fusion will most probably reach commercial viability.

Contrary to prevailing perceptions, current fusion technology is not inherently "clean," because neutrons escape, making surrounding materials and the reactor radioactive. But other fusion technologies include advanced fuel cycles that perhaps would be neutron-poor or neutron-free—allowing fusion to become a clean source of energy.

Hybrid reactors might someday combine fusion and fission. Neutrons from fusion reactions would be absorbed by a blanket of fissionable material, which in turn would convert natural uranium or other elements into fissile material.

Another emerging technology, electric breeding, could also be important. An electric breeder would speed protons into a target of uranium nuclei or other elements. The resulting material could be used in fission reactors.

Finally, second-generation chemical processing not only could separate plutonium and uranium from the fission products but also could remove certain burdensome long-lived isotopes during the process. This reprocessing technology could lead to second-generation final waste disposal, requiring much less space when these isotopes are separated out. Opening a final waste disposal every other year or more would be unnecessary.

Although the construction of some 2,000 additional reactors worldwide is technologically possible, a wide chasm exists between such feasibility and political and societal constraints. The World Commission on Environment and Development issued a report in 1987 entitled "Our Common Future," which recognized nuclear power's potential but was hesitant to embrace a 2,500-reactor scenario without certain legally binding qualifications. The Brundtland report, named after Gro Harlem Brundtland, who chaired the commission, stipulated that there be early-notification procedures in case of a nuclear accident or the release of excessive

Figure 9.6 TARAPUR REACTOR, a BWR in Maharashtra, India, was connected to the global electric grid in 1969 and currently provides 300 megawatts of electricity. Tarapur is the site of one of the world's four operational chemical-reprocessing plants.

amounts of radioactivity, emergency response training, regulations for the transport of radioactive materials, standardization of operator training and licensing procedures, enactment of reactor operation and safety rules, reports of routine and accidental discharges, site selection criteria, specifications for waste repositories, and procedures for decontamination and dismantling. Significantly, these concerns are not technological ones but rather institutional.

Many of these concerns have been addressed by international organizations — the IAEA, Euratom and the World Association of Nuclear Operators, among them. But unwilling countries cannot be forced to adhere to these agencies' standards, and quality control, stringent operation and sound management are difficult to ensure without enforcement power. As Peter W. Beck, formerly of Shell London, wrote: "Many of us may well have missed the most important lesson of Chernobyl: the international nature of nuclear safety. In other words, the paramount need for safe design and operation everywhere is the concern of everyone irrespective of frontiers or distance."

Unfortunately, the international consensus that nuclear power demands by its very nature is being obscured by national debates that center around the reprocessing and waste disposal issues. These concerns may turn out to be the principal obstacle for nuclear power.

Efforts in the past 20 years to develop civilian (as opposed to government-operated) reprocessing facilities have had only limited success. La Hague in France and Sellafield in England are models of such civilian-run operations, although in each case the government maintains a presence. Similar efforts have not been successful in the U.S. or in West Germany. In thee 1970's the Carter administration passed the Nuclear Non-Proliferation Act, which called for the termination of all civilian reprocessing activities, with an eye to having such processing halted in other countries as well. Wackersdorf in West Germany stopped construction last year, forcing the utilities there to turn to France and Britain for processing.

Controversy continues to erupt over disposal, given that all reactors generate some amount of waste. But there are solutions. Karl P. Cohen, formerly of the General Electric Company, proposed that the U.S. adopt an intermediate solution. He suggested that spent fuel be stored on site for at least 100 years. Storage, however, will eventually be a global problem. There should be an international institution—preferably IAEA or one under its auspices—to construct and operate one or more temporary retrievable storage facilities. The capacity of such facilities should initially be some 400,000 tons of heavy metal. These facilities could be located on islands or peninsulas. A variation of the idea was proposed by Cesare Marchetti in the early 1970's and again by myself in 1976.

International storage facilities offer several advantages. They encourage the development of global institutions that would be immune to national politics. Such facilities would allow the nuclear-power industry the time it needs to develop scientific, technological and institutional final waste disposal methods. Access to these facilities would give countries that steered clear of nuclear power because of the waste issue a chance to develop nuclear energy. The sites could also play a key role in disassembling nuclear weapons and ensuring nonproliferation of nuclear material.

If their installation and operation were successful, such sites could also lay the foundation for final waste disposal. And if sites were chosen through global consensus, fuel-reprocessing and possibly other nuclear facilities could be constructed and operated there as well.

This idea of the international regulation of nuclear power was first presented in 1946 by the Acheson-Lilienthal-Baruch plan. These early fathers of nuclear energy had a clear understanding of the potential power of nuclear energy (the one trillion factor) and were driven by their concern that nuclear weapons might proliferate through the spread of fissionable material.

But the Acheson-Lilienthal-Baruch plan was never realized. Instead the 1954 Atomic Energy Act initiated its Atoms for Peace Program. It was followed in 1970 by the Nuclear Non-Proliferation Treaty, which has been signed by 140 nations, establishing the international safeguard system still successfully administered by the IAEA.

The IAEA safeguard system consists of three elements: material accountability, surveillance and containment—within a framework of objectivity, rationality and efficiency. There is broad agreement that quantitative measurement comes closest to fulfilling these criteria. Therefore, material account-

ability is paramount among the three. For the majority of nuclear reactors, material accountability leads to the identification and tallying of specific items with relative ease. But such monitoring is more challenging with reprocessing, enrichment and fuel-fabrication facilities; surveillance measures and resident inspectors become increasingly critical at these sites.

It is important to remember that the proliferation of nuclear weapons is not necessarily related to civilian uses of nuclear power. In fact, a crude centrifuge enrichment device can provide the same access to such armaments. And there is a great distinction between a single crude nuclear device and an operational nuclear armory. Ensuring nonproliferation is hardly a straightforward task.

Today's nonproliferation regime—that is, the web of technical, institutional and political measures already in place—works rather well, that is, I would guardedly give it a positive evaluation. Therefore, I do not think an expansion of nuclear capability from 400 GWe to 2 TWe would contribute to proliferation—provided certain standards are maintained. It should not be too difficult to apply item accountability to new reactors. And because the few enrichment facilities that currently exist are effectively safeguarded, increasing the number of enrichment facilities and fuel-fabrication facilities by a factor of five should pose no hardship.

No mention of costs was made in this article expressly. Some would say that nuclear power has declined in certain countries because of the excessive expenses and long construction times associated with new plants; yet costs are established not on paper but by market demand. Real obstacles to the development of nuclear power have been regulatory and licensing uncertainties and public opposition. Construction delays of eight years or so have sometimes doubled the cost of a reactor.

Nuclear power must play a significant role in a sustainable future, but it must be carefully administered. Safety standards, design features, waste storage and the entire fuel cycle must be under the jurisdiction of an international regulatory agency. Only then can the full potential of nuclear power be realized. In view of worldwide population growth and its accompanying problems, I sincerely doubt that turning backward will allow us to master the future. Instead we have to go forward.

Energy from the Sun

*Various forms of solar energy, including wind and biomass,
offer environmentally benign ways to generate electricity and make
fuels. Some technologies will be cost-competitive before the year 2000.*

. . .

Carl J. Weinberg and Robert H. Williams

Worries about urban air pollution, acid rain, oil spills, nuclear risks and global warming are prompting a reexamination of alternatives to coal, oil and nuclear power. Although alternative energy sources are not universally pollution-free, there is a wide range of options that are far less environmentally damaging than conventional energy supplies. The most promising technologies harness the sun's energy. Here we examine a selected set of solar technologies for producing electricity and fuels and discuss strategies for realizing their potential.

Several solar technologies and the industrial infrastructures needed to exploit them are advancing rapidly. Electricity from wind, solar-thermal and biomass technologies is likely to be cost-competitive in the 1990's; electricity from photovoltaics and liquid fuels from biomass should be so by the turn of

the century. While it is premature to predict which of these will dominate, it is clear that the solar energy supply will be diversified in both technology and scale and will have marked regional variations. These characteristics imply the need for new approaches for both introducing and managing solar technologies.

Hydropower is the most highly utilized solar energy source. In this case, the sun's role is indirect. Sunlight evaporates water, which later falls as rain; rainwater flows into rivers and turns generator turbines as it returns to the sea. In 1987 hydropower accounted for 17 percent of electricity production in industrialized countries and 31 percent in developing countries. The World Energy Conference estimated that the amount of hydropower that could be exploited commercially is nearly five times that now being generated. The hydroelectric potential in developing countries is particularly large—nearly 10 times the amount already developed.

It is unlikely, however, that all of this potential will be developed, as hydroelectric projects increasingly become targets of environmental concern. Frequently discussed problems include the loss of large land areas to hydroelectric facilities, the potential for catastrophic dam failures and various

Figure 10.1 WIND FARMS at Altamont Pass, Calf., contain 7,500 wind turbines owned and operated by independent companies who sell the electricity to Pacific Gas & Electric. During the 1980's mass-production techniques and improved deployment and operating strategies cut the cost of electricity from wind tenfold. Modular solar energy systems are especially amenable to such cost cutting.

health and ecological worries. Yet if electricity is used wisely so that the growth in demand is slow, even the limited untapped resources defined by environmental constraints could meet a significant fraction of future electricity needs.

A mong other solar electricity options, wind power is closest to being economically competitive. Wind is solar power that has already been converted into mechanical power, so further conversion to electricity can be accomplished efficiently.

During the 1980's some 1,660 megawatts of wind-electric capacity was installed worldwide (a megawatt is one million watts). Of the total, 85 percent is in California, mostly at Altamont Pass in territory served by the Pacific Gas & Electric Company (PG&E), where there are now about 7,500 wind turbines (see Figure 10.1). The California wind boom resulted largely from favorable tax policies and high prices paid by utilities for wind-generated electricity in the mid-1980's. These incentives have been discontinued, but wind power continues to grow in California, albeit at a slower rate.

The Altamont wind farms have sometimes been referred to disparagingly as tax-shelter energy investments. Indeed, difficulties arose in the early years. Tax incentives encouraged rapid construction of wind machines whose designs had not been rigorously tested, and failures were common. Today most of the problems have been resolved, and remarkable improvement in wind-power economics has been made. Since 1981 the cost of wind-generated electricity has dropped nearly an order of magnitude, to less than seven cents per kilowatt-hour (for comparison, electricity from a new coal-fired power plant costs about five cents per kilowatt-hour in the U.S.).

Few of the cost reductions can be traced to improved technology. Except for lightweight composite-material blades and microprocessor-controlled turbines, commercial wind turbines at Altamont incorporate no substantial aerodynamic or design innovations over those built 50 years ago. The reduced cost of wind power stems mainly from organizational learning, which involves standardizing procedures. At the factory, manufacturers learned mass-production techniques; in the field, workers learned to site machines more effectively and to schedule maintenance at times of low wind.

For large-scale (500 to 1,000 megawatts per unit) fossil-fuel or nuclear-power plants, the construction of a single plant is so complicated and time-con-

suming that advances along the learning curve are slow at best, and creating standardized designs is difficult. In contrast, the small unit size (50 to 300 kilowatts) and relative simplicity of wind and many other solar technologies make it possible to mass-produce identical units. The time required from initial design to operation is so short that needed improvements can be determined by field testing and quickly incorporated into modified designs.

New, sophisticated wind turbine technologies promise further savings. PG&E is involved in a five-year cooperative effort with the Electric Power Research Institute (EPRI) in Palo Alto and U.S. Windpower in Livermore to develop, build and test prototypes of a 300-kilowatt variable-speed wind turbine. These turbines will have blades and sophisticated electronic controls that allow the rotor to turn at optimal speed under a wide range of wind conditions, thereby increasing wind energy capture. These innovations offer other benefits as well, including reduced material fatigue and lower maintenance costs.

Further gains in wind-power efficiency could be achieved with advanced airfoils, "smart" electronic controls that adjust system-operating parameters based on wind characteristics, and continued development of advanced materials that yield lighter, stronger components. Taken together, these new technologies should make wind power cost-competitive. The U.S. Department of Energy and industry analysts project that during the next 20 years the costs of wind electricity at sites with moderate wind resources could fall to 3.5 cents per kilowatt-hour.

The economics of wind power depend on value as well as cost. One might think that, because wind is intermittent, wind electricity without storage would have a value no greater than the fuel and operating costs saved by not having to produce electricity in conventional power plants, but wind power also has the value of reducing the need to build conventional generating capacity.

The value of wind power depends both on conditions at the site and on the nature of the local utility's demand profile. To the extent that wind power is available at the time of the utility's peak demand, it is especially valuable; peaking power is the most costly to provide. In northern California, wind output, averaged over the two major wind sites (Altamont and Solano), is available at 50 percent of rated generating capacity during this period. For comparison, the availability of fossil-fuel power plants is 80 to 90 percent of rated capacity.

Wind energy is relatively clean, and most of its

problems have been solved. Noise was one concern, but modern turbines make little sound beyond the rush of the wind. Steel blades can interfere with television reception, but this has not been a problem with the relatively small Altamont turbines, whose rotors are made mostly of fiberglass or wood and epoxy. At Altamont, bird kills are a possible problem that is being investigated. Perhaps the most serious problem is aesthetic: some people do not want to see windmills on the landscape.

About 90 percent of the wind power potential of the U.S. is in 12 contiguous states (see Figure 10.2), where large-scale ranching and grain production are major industries. Wind power could be a good neighbor to such agricultural activities. Experience shows that the value of ranchland increases rapidly when it is converted to wind farms. At Altamont, prices rose from $400 to $2,000 per acre, reflecting royalties paid to landowners for use of the land. Ranchers lost only about 5 percent of the grazing area at Altamont, where cattle graze around the wind machines.

Wind resources in the U.S. are concentrated in states where potential generation would greatly exceed local demand. Therefore, it is important to improve electricity networks and reduce transmission costs to make it economical for these states to become major electricity exporters.

A promising energy technology for sunny areas, known as solar thermal electric generation, employs reflective solar collectors that track the sun and concentrate its heat and light. The concentrated sunlight heats a fluid, which is used in a power-generating cycle.

Between 1984 and 1988 the LUZ Corporation based in Los Angeles installed several commercial solar-thermal electric plants having a total generating capacity of 275 megawatts in California's Mojave Desert. LUZ has 80 megawatts of capacity under construction and will install another 300 megawatts at Harper Lake in southern California by 1994. The LUZ system uses mirrors mounted in parabolic troughs to focus sunlight on oil-carrying receiver pipes. The oil is heated as it circulates through the pipe and is used to create steam that drives a turbine generator (see Figure 10.3). A natural gas burner can be used to augment the solar heat as needed.

Like wind power, solar-thermal electric technol-

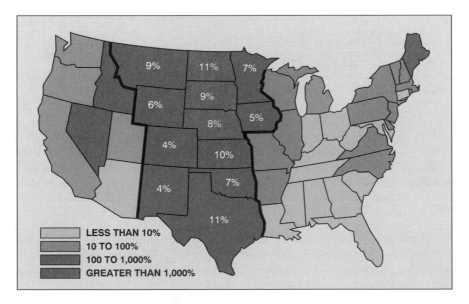

Figure 10.2 U.S. ELECTRICITY DEMAND could be satisfied by four million 500-kilowatt wind turbines spaced half a kilometer apart over 10 percent of the U.S. where wind is favorable. Colors show wind-electric potentials as a percentage of in-state electricity generation. Roughly 90 percent of the U.S. potential for wind is in 12 states, indicated by the bold margin. The numbers refer to the percentages of total U.S. wind-electric potential for these states. Wind-generated electricity in these states could far exceed local demand; large amounts could be exported or used to make hydrogen.

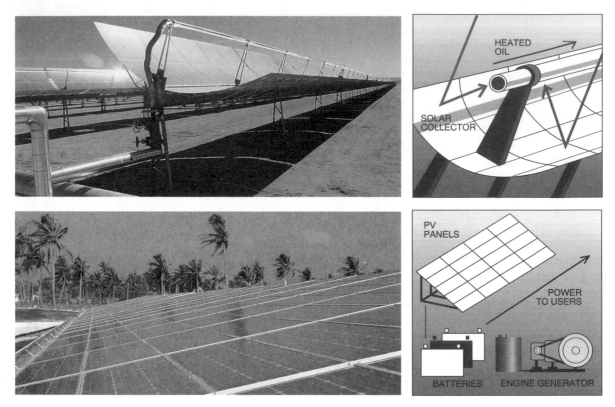

Figure 10.3 SOLAR-THERMAL ELECTRIC plant built in the Mojave Desert by the LUZ Corporation produces electricity that is sold to Southern California Edison. Concentrated sunlight heats oil in the pipes; heat from the oil generates steam that drives a turbine generator (*top*). A natural gas burner provides supplemental heat during periods of high demand or low sunlight. Photovoltaic cells that convert sunlight directly into electricity are now cost-effective in some isolated locations, such as Coconut Island off the Australian coast. Batteries and a diesel engine generator provide backup power (*bottom*).

ogy involves modular construction and offers the economies of mass production. Each of the 80-megawatt plants at Harper Lake will consist of 852 100-meter-long, independently operated solar-collector assemblies. Construction time for the first of these plants, which went into operation in 1989, was nine months, far less than the six to 12 years typical for conventional central-station power plants. Improved engineering, manufacturing and construction techniques have reduced electricity costs from 23 cents per kilowatt-hour for the first LUZ plant to 10 cents per kilowatt-hour for plants being built today.

An alternative solar-thermal electric concept involves sun-tracking mirrors that focus concentrated sunlight onto a central receiver, reaching much higher temperatures than is feasible with parabolic troughs. This approach is especially well suited for use with promising solar-thermal storage technologies. In the 1970's a number of pilot plants were built, including the 10-megawatt Solar One plant at Daggett, Calif., and a five-megawatt unit at Almeria, Spain. The major central-receiver project at present is the European Phoebus project, a 30-megawatt plant to be built in Jordan. If successful, Phoebus could reestablish this approach as a viable option.

Photovoltaic (PV) electricity is produced directly from solar energy when photons (individual particles of light) absorbed in a semiconductor create an electric current. PV is the quintessential energy source, creating electricity with no pollution, no noise and often no moving parts. Photovoltaic systems need minimal maintenance and no water and so are well suited to remote or arid regions. They can also operate on any scale, from multiwatt portable modules for remote communications and instrumentation to multimegawatt power plants covering millions of square meters. This size flexibility makes it possible to locate PV systems near users, where the produced electricity is more valuable than at a central station. Small PV systems therefore are potentially cost-effective even in some relatively cloudy or high-latitude areas, where solar energy might seem impractical.

PV power is unlikely to be limited by land-use constraints. A 12 percent efficient, 40-square-meter array mounted on a south-facing rooftop in an area of average solar radiation in the U.S. could produce about as much electricity as is consumed by a typical U.S. household. The amount of PV electricity equivalent to total U.S. electricity generation could be produced on a collector field covering 34,000 square kilometers, or less than .37 percent of the U.S. land area.

The rapid decline in PV costs (from $60 per kilowatt-hour in 1970 to $1 in 1980 and to 20 to 30 cents today) and the development of niche markets have increased PV demand at a rate of 25 percent a year. Annual worldwide sales now exceed 40 megawatts of peak capacity. The present cost is still about five times the cost of electricity from conventional sources, but progress has been more rapid than anticipated.

Solar-cell efficiencies in the laboratory have improved from 16 to 18 percent in the mid-1970's to today's 28.5 percent for a point-contact crystalline silicon cell and 35 percent for a gallium arsenide-gallium antimonide stacked junction cell (a cell with two layers that absorb different parts of the solar spectrum).

A promising new class of solar cells based on thin films of semiconductor material is being developed. Although these cells generally have lower efficiencies (the highest yet achieved in the laboratory is less than 16 percent), they have the potential for very low cost, perhaps as low as one-tenth the current PV module market price. Thin films are especially amenable to cost-cutting, mass-production techniques and require only tiny amounts of active material. The films are from only one to two microns thick, one-fiftieth the thickness of a human hair.

Further improvements are needed before solar cells will be competitive in the bulk power market. Crystalline silicon is efficient and reliable, but its production costs remain high. Thin-film technologies have yet to demonstrate adequate levels of efficiency and reliability in commercial products. And as the cost of PV modules decreases, more attention must be paid to reducing the costs of other PV system components, which already account for half of total PV costs.

To examine the prospects for large-scale PV electricity generation, PG&E created a government-industry partnership called the Photovoltaics for Utility Scale Applications (PVUSA) project. The project is designed to bridge the gap between PV research and development and commercial implementation. A variety of systems are being tested in order to provide comparative data on reliability and performance, operation and maintenance costs as well as innovative systems designs.

Similar projects are under way around the globe. In Japan all major utilities are involved in PV projects, such as the Rokko Island test facility, where 100 residential-scale PV systems connected to the power grid are being tested. The largest West German utility, Rheinisch Westfälisches Electrizitätswerk, is evaluating several PV technologies at an installation that will eventually generate one megawatt (peak). The governments of Italy and Spain have installed and are funding freestanding systems for homes and remote areas. Italy is also planning larger systems to generate up to three megawatts.

Concerted efforts also are needed to promote the growth of PV manufacturing industries. One way to do this is to target niche markets in which the value of electricity is particularly high. Consumer electronic products (solar calculators, watches and so on), for example, raised amorphous silicon solar cells from a laboratory curiosity to a viable commercial technology. These products now account for nearly 40 percent of PV sales worldwide.

PV is already cost-effective in applications far from existing power lines, such as remote residencies, research stations and military and communications facilities. Several U.S. utilities are using PV systems for small-scale operations. The Georgia Power Company found that a $3,000 PV remote lighting system eliminated the need for a $35,000 electric grid extension.

Regions of developing countries where rural electrification is embryonic are important early markets for PV systems. Extending power lines from centralized sources to rural areas is often not yet economical, and so decentralized power sources such as PV are a promising alternative. On a lifetime-cost basis, PV is now cost-effective compared with diesel generators at capacities below 20 kilowatts. In India there are from four to five million diesel-powered water pumps, each consuming about 3.5 kilowatts. This market alone could support annual PV sales of perhaps 1,000 peaks megawatts, 25 times current global sales. Niche markets account for only a minuscule share of the global energy market, but they would help establish a viable PV industry that could begin competing in bulk power markets around the turn of the century.

Many believe that because of their variability, wind and direct solar energy can meet only a small fraction of total electricity demand without a substantial amount of energy storage. The need for storage depends on the temporal relation between the availability of energy from renewable sources and the utility's demand. The amount of storage required could be surprisingly small. In California demand peaks daily in the late afternoon and yearly

in the summer, when the potential supply from wind and direct solar resources is greatest. It therefore would be feasible to meet as much as half of the peak demand and one-third of the total electric energy with wind and solar sources by adding only a modest amount of storage (see Figure 10.4).

Even so, direct solar and wind energy could make a greater contribution if more storage were available. It is therefore important to improve storage technologies in parallel with improvements in solar electric-generating technologies. Solar-thermal electric systems are readily adapted to high-temperature thermal storage. Today this is potentially more economical than batteries or other nonthermal storage, especially when the fluid that transports heat from the collector is also the heat storage medium. Thermal storage makes it possible for a solar-thermal electric plant to provide power like a base-load fossil-fuel power plant. Many studies indicate that central-receiver thermal-electric systems with thermal storage could be competitive with fossil-fuel systems.

Another approach is to add a low-cost fossil-fuel system whose output can easily be adjusted to com-

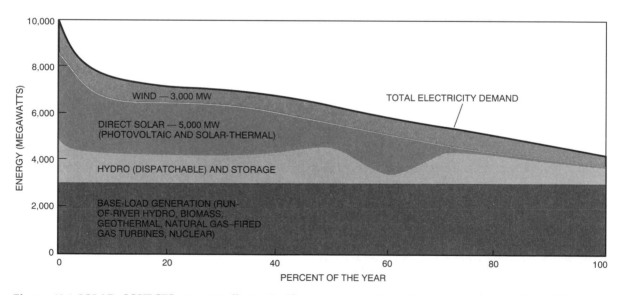

Figure 10.4 SOLAR SOURCES can contribute significantly to electricity supplies, as illustrated for a hypothetical utility in California. Each point on the top curve shows the percentage of the time that utility demand exceeds the indicated value. The 1 percent of the time (88 hours a year) when the load exceeds 90 percent of peak demand accounts for 90 percent of the annual probability that the utility

cannot meet its load because of system failures. Wind and direct solar energy can satisfy half of peak demand and a third of total electric energy. Hydroelectric and biomass sources could increase the solar contribution. The dip in the hydropower curve (green) results from the availability of solar energy on weekends, when demand is low.

pensate for variations in solar supply. A solar-natural gas "hybrid" in particular would be one of the most environmentally benign ways of using fossil fuels to generate power.

At very low incremental cost, LUZ added a natural gas-firing capability to its solar-thermal electric system. The supplemental natural gas heat source gave LUZ the capability to maximize the value of its output to the utility, by ensuring that power is available when needed.

The solar-natural gas hybrid concept is also applicable to wind and photovoltaics. A new generation of efficient, inexpensive gas turbines could be used in these hybrids. Although the extra cost would be greater than for the gas capacity in the LUZ hybrid, these gas turbines would be only about one-third as costly as coal-fired steam-electric capacity.

Hybrid "miniutility" systems containing PV, batteries and diesel generators already provide reliable power for some remote applications. A miniutility system of this kind serves a community of some 100 people on Coconut Island in the Torres Strait between Australia and New Guinea. Similar systems are being considered for Africa 1000, a project to provide electricity to 1,000 villages there, and for the Australian outback.

Some solar energy options have "built-in" storage. In biomass — green plant matter created in photosynthesis — solar energy is stored as chemical energy that can be recovered by burning the plants.

Biomass offers a number of advantages. Unlike fossil fuels, biomass is available over much of the earth's surface. It generally contains less than .1 percent sulfur and 3 to 5 percent ash, compared with 2 to 3 percent and 10 to 15 percent, respectively, for bituminous coal. If biomass is produced at a sustainable rate, the carbon dioxide released when biomass is processed and burned exactly balances the carbon dioxide consumed during photosynthesis. Bioenergy would make no net contribution to the carbon dioxide in the atmosphere, and so it would not contribute to global warming (see Figure 10.5).

Biomass is widely used to generate electricity and heat in the forest-products industries. Wood wastes from the production process are used as fuel for steam-turbine cogeneration systems. This approach is economical only where low-cost biomass fuel is readily available. Steam turbines are relatively expensive and inefficient at the modest scales that are practical for biomass (less than 100 megawatts). At larger scales, fuel transportation costs become prohibitive because biomass resources are distributed thinly over large areas.

The gas turbine offers a promising means of generating cost-competitive biomass power, even using more costly biomass fuels such as logging waste or plants grown on plantations. In gas turbines, gaseous fuel is burned, and the hot combustion products are directed to a turbine that generates electricity. In addition, the hot turbine-exhaust gases can be used to produce steam, which can be utilized for industrial applications or for additional power generation. Gas turbines are inherently simpler and cheaper than conventional steam turbines. And whereas the latter have shown no improvements in efficiency since the late 1950's, gas turbines have improved continuously.

The most promising way to use biomass in gas turbines is to gasify it with air and steam at high pressures and clean the gas of impurities that might damage the turbine blades before burning it. Gasification and power generation would occur at the same facility to maximize efficiency. Such integrated gasifier-gas turbine technology is now being developed for coal, and it could easily be adapted to biomass. In fact, the technology probably could be commercialized more quickly and cheaply for biomass than for coal because biomass is easier to gasify and generally contains little sulfur. Preliminary estimates by one of us (Williams) and colleagues at Princeton University indicate that a biomass gasifier-gas turbine could compete in cost with conventional coal, nuclear and hydroelectric power in both industrialized and developing countries.

The most promising near-term applications for biomass-powered gas turbines are in industries where large quantities of biomass residues are readily available, such as the cane sugar and alcohol industries. Gas turbines fueled with sugarcane residues could generate far more electricity than sugar factories or alcohol distilleries need. At Brazilian distilleries the economic benefits of electricity co-production could make alcohol competitive with current, low oil prices. At the present level of cane production, gasifier-gas turbine systems could produce about half as much power as is now generated by all sources in the 80 developing countries that produce cane.

Although electricity production will probably dominate initial applications of solar energy, the production of liquid and gaseous fuels also will

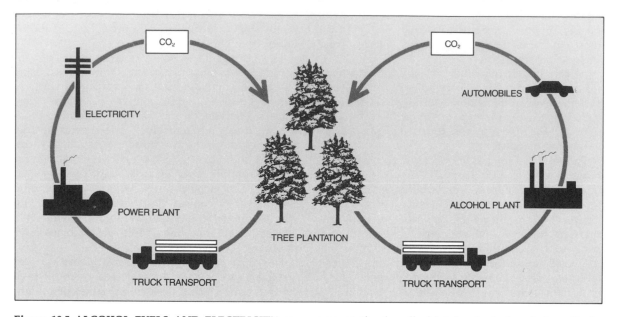

Figure 10.5 ALCOHOL FUELS AND ELECTRICITY can be produced from biomass (wood or other plant matter). Burning biomass releases carbon dioxide, but if biomass is grown sustainably, the amount released is balanced by the amount of carbon dioxide taken in during photosynthesis. Biomass energy, properly managed, would not contribute to greenhouse warming.

be important in the 21st century. Interest in synthetic fuels flared briefly during the 1970's, waned with declining oil prices in the 1980's and recently has reemerged, this time driven primarily by environmental concerns.

Methanol is one much discussed alternative fuel for transportation (see Chapter 5, "Energy for Motor Vehicles," by Deborah L. Bleviss and Peter Walzer). Although methanol would create less local air pollution than gasoline, a shift to fossil fuel-based methanol poses climate risks. Methanol would probably be produced from natural gas at first, but as gas supplies tighten, suppliers might switch to coal, which is much more abundant, as the raw material. Coal-based methanol would exacerbate global warming because it releases twice as much carbon dioxide into the atmosphere as does gasoline.

This problem could be avoided by using methanol produced from woody biomass that is grown sustainably. The biomass would be gasified and synthesized into methanol by processes similar to those being developed for coal. Coal-based methanol would be less costly if, as currently contemplated, the coal facilities are large units that cost a billion dollars or more. The trend in energy conversion, however, is toward smaller, more modular and less financially risky units. At the much smaller scales needed for biomass, biomass-derived methanol would be cheaper than coal-derived methanol (see Figure 10.6).

One alternative to methanol is ethanol produced by the fermentation of biomass-derived sugars. Ethanol from sugarcane has been produced on large scales in Brazil; in the U.S., modest quantities of ethanol produced from corn are used as a gasoline extender. Corn-based ethanol is relatively expensive because growing corn is costly; however, technology is being developed for making fermentable sugars out of low-cost woody feedstocks using enzymes. Researchers at the Solar Energy Research Institute (SERI) in Golden, Colo., think that by the year 2000 ethanol produced from such inexpensive sources could be competitive with gasoline.

Total biomass production will ultimately be limited by land and water availability because of the low efficiency of photosynthesis and the large water requirements for growing plants. Nevertheless, biomass can play a significant role in the energy economy if energy efficiency is emphasized. In the U.S.,

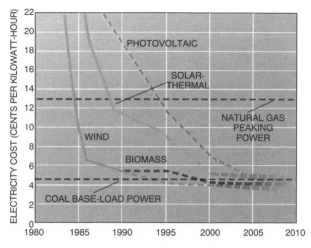

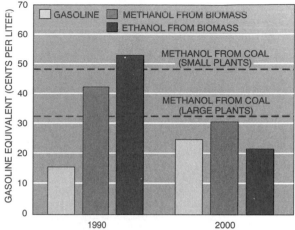

Figure 10.6 SOLAR ELECTRICITY COSTS fell sharply in the 1980's and will continue to fall as technology improves and experience is gained. Electricity costs (*left*) are based on actual (*solid*) or projected (*broken*) costs for bulk power generation. Costs are also shown for new coal-fired base-load and natural gas-fired peaking plants. Biomass-de-rived methanol would cost more than methanol from large coal-conversion plants, but if coal plants were the same size as biomass plants, methanol from biomass would be cheaper. Ethanol derived from wood could be cost-competitive with gasoline by 2000 (*right*).

potential biomass supplies could replace all oil now used in light-duty vehicles and coal now burned for power generation, provided vehicle fuel economy is doubled and efficient gasifier-gas turbine units are used for power. In this way, national carbon dioxide emissions could be cut in half (see Figure 10.7).

Hydrogen produced electrolytically, using solar electricity to split water into its constituent elements, is a clean fuel that could be used for transportation and heating and eventually for producing electricity and by-product heat in highly efficient fuel cells (see Chapter 8, "Energy from Fossil Fuels," by William Fulkerson, Roddie R. Judkins and Manoj K. Sanghvi).

Conversion to hydrogen provides a convenient means of storing intermittent solar energy. Moreover, because in principle it costs less (on a lifetime-cost basis) to transport hydrogen by pipeline than to transmit electricity by wire, hydrogen facilities can be located where production is cheapest, even if such sites are far from where the hydrogen will be used (see Figure 10.8).

Advances in solar technologies during the next two decades should make it feasible to produce electrolytic hydrogen from these sources at a cost to consumers that is about twice the present gasoline price in the U.S., on an energy-equivalent basis. This is well below prices consumers now pay for gasoline in Europe and Japan, where gasoline taxes are high.

The primary appeal of solar-derived hydrogen compared with fossil fuels is its environmental friendliness. When hydrogen is burned, it turns back into water. It produces no carbon monoxide, carbon dioxide, sulfur dioxide, hydrocarbons or particulate matter. The only pollutants are oxides of nitrogen, which can be reduced to very low levels.

Solar hydrogen will also be an attractive fuel where or when land and water constraints limit the possibilities of fuels from biomass. For example, the amount of PV hydrogen equivalent to the total world fossil-fuel consumption could be produced on 500,000 square kilometers, less than 2 percent of the world's desert area. In fact, sunny deserts are promising sites for hydrogen production, because water requirements for electrolysis are modest — for PV hydrogen, equivalent to only two to three centimeters of rainfall per year. Solar-derived hydrogen makes it possible for solar energy sources to play a larger role in fuel production than is possible with biomass alone.

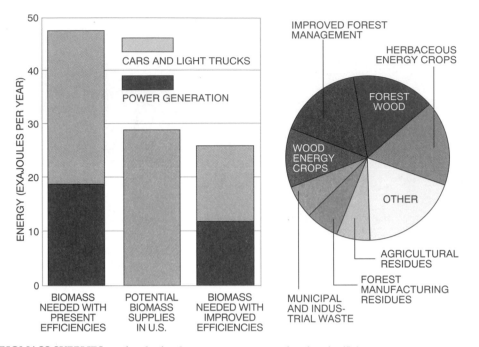

Figure 10.7 BIOMASS SUPPLIES can be obtained economically from a number of sources in the U.S. They could displace coal for electricity generation and oil for light vehicles, which in 1987 together accounted for 49 percent of fossil-fuel carbon dioxide emissions in the U.S. At present levels of efficiency, biomass could supply two-thirds of the combined energy need. Doubling vehicle fuel economy and switching to advanced gas turbines for power would enable biomass to meet the entire need.

Solar energy technologies are advancing quickly, and the prospect for further gains is auspicious. Specific policies needed to promote large-scale development of solar energy will vary from one part of the globe to another, but some general guidelines for policy-making can be outlined.

Efficient energy use will make solar energy more attractive by extracting more useful services from solar sources and by decreasing the capital needed to provide energy services. Therefore, promoting efficient energy use should be a key element of solar energy policy.

The rules of the present energy economy were established to favor systems now in place. Not surprisingly, the rules tend to be biased against solar energy. A major challenge to policymakers is to overcome these biases.

One set of biases involves the taxes and subsidies that encourage the exploitation of fossil fuels and favor operating costs over long-term investments. Such biases should be eliminated, or if that is not politically practical, at least the system of taxes and subsidies should be more balanced.

Another set of biases arises because present energy prices usually do not reflect many of the external social costs of energy production and use, including the dangers from air pollution and nuclear risks and the economic, ecological and human health costs of global climate change. Such costs tend to be lower for solar sources. Policies that take them into account would render solar energy more competitive.

Despite the dramatic progress that has been made during the past decade in advancing solar energy technologies and reducing their costs, most are still not competitive with conventional energy on a direct-cost basis. Further research and development is needed, although simply increasing R&D budgets is not enough. An important lesson of the past decade is that technology developed in the laboratory and then just passed to a commercial developer does not guarantee market acceptance. A better approach is

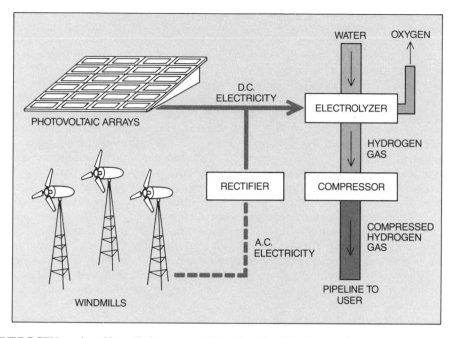

Figure 10.8 HYDROGEN produced by splitting water with photovoltaic or wind electricity is a clean fuel that stores solar energy in a chemical form. Transporting hydrogen is in principle cheaper than transmitting electricity, so converting to hydrogen can be an attractive means of bringing solar energy to major demand centers.

to promote cooperation among researchers, producers and potential users in consortia or joint ventures, accelerating learning in the laboratory, the factory and the field.

Government support should both address basic and applied science and technology issues of broad generic importance and foster the industrial capability for developing, deploying and managing solar energy systems. Developmental projects should be designed to advance the targeted technologies and to develop industrial capability, with costs shared among the government, producers and potential users.

A major challenge is the creation of an industrial infrastructure for solar energy. The most appropriate approaches will vary by technology and by region. The dramatic cost reductions for wind and solar-thermal power in California resulted from favorable tax policies and power-purchase agreements established by the state. Such policies created a flow of revenue that funded the product improvements and organizational learning, causing costs to decline.

Questions of economic efficiency arise when incentives are used to promote new technologies. Nevertheless, to the extent that incentives are crafted to reflect external social costs not captured by market prices, they offer a sound approach to nurturing solar industry development.

Another approach is to develop high-value niche markets before bulk energy markets. The solar energy industry is relatively small, and so it hesitates to expand production capacity without some assurance that the market will grow. Utilities and governments could help by identifying niche markets and by purchasing solar energy systems for markets under their purview.

In some instances, the barriers to market development are institutional rather than economic. There are substantial biomass cogeneration markets that can be exploited cost-effectively in the near term, using cheap biomass residues from industrial activities (such as the pulp and paper and sugarcane industries). But in many parts of the world such markets cannot readily be developed because utilities are unwilling to purchase the excess electricity or to offer backup power at fair prices.

A final policy issue concerns industrial structures

for solar energy. The nature of solar resources and technologies implies that although the solar energy industries will probably have some structural features quite different from those of today's energy industries, present industries have unique capabilities that could help solar energy reach its full potential.

The modular nature of solar technologies suggests that companies that produce solar equipment should be modeled less after those that make large-scale energy-production technologies and more after successful companies in other industries that offer products amenable to cost-cutting, mass-production techniques. The diversity of solar supplies, in technology and scale, also poses major challenges for routine energy-supply management.

Consider electric utilities. Today they produce power in centralized plants and send it to their customers through large transmission and distribution networks. The present system is built around the notions of centralization and interconnectedness, whereas solar technologies often offer the greatest value in dispersed, grid-connected or stand-alone systems. Nevertheless, the electric utility is the logical institution for managing solar electricity.

For solar energy systems, supplies will have to be organized into a reliable, cost-effective system, whether the electricity is produced by utilities or independent power producers, or both. Even for stand-alone systems, utilities can play major roles. Today such systems are often installed by groups or agencies other than utilities because of the view of both utilities and government that utilities should be involved only with "electricity by wire." Utilities, however, should also offer electricity without wire as a service option, because utility expertise is invaluable in providing monitoring, maintenance and general servicing on a continuing basis.

The successful involvement of existing energy companies in managing solar energy requires that they rethink their basic structures. Their task will be more difficult than at present because of the diversity and variability of solar energy sources, as well as the challenges of managing storage technologies and reorganizing conventional sources to complement solar sources in the most efficient manner. Utilities must place less emphasis on centralization and interconnectedness derived from economies of scale and more on their ability to offer the economies of scope needed to manage solar resources.

Experience from the 1980's shows that public policy can accelerate market acceptance of solar energy. Framing effective policies for hastening the transition to solar energy is one important, responsible strategy for addressing growing environmental concerns.

Energy in Transition

*The era of cheap and convenient sources of energy
is coming to an end. A transition to more expensive
but less polluting sources must now be managed.*

. . .

John P. Holdren

As the foregoing chapters make clear, civilization is not running out of energy resources in an absolute sense, nor is it running out of technological options for transforming these resources into the particular forms that our patterns of energy use require. We are, however, running out of the cheap oil and natural gas that powered much of the growth of modern industrialized societies, out of environmental capacity to absorb the impacts of burning coal, and out of public tolerance for the risks of nuclear fission. We seem to be lacking as well the commitment to make coal cleaner and fission safer, the money and endurance needed to develop long-term alternatives, the astuteness to embrace energy efficiency on the scale demanded and the consensus needed to fashion any coherent strategy at all.

These deficiencies suggest that civilization has entered a fundamental transition in the nature of the energy-society interaction without any collective recognition of the transition's character or its implications for human well-being. The transition is from convenient but ultimately scarce energy resources to less convenient but more abundant ones, from a direct and positive connection between energy and economic well-being to a complicated and multidimensional one, and from localized pockets of pollution and hazard to impacts that are regional and even global in scope.

The subject is also being transformed from one of limited political interest within nations to a focus of major political contention between them, from an issue dominated by decisions and concerns of the Western world to one in which the problems and prospects of all regions are inextricably linked (see Figure 11.1), and from one of concern to only a small group of technologies and managers to one where the values and actions of every citizen matter.

Understanding this transition requires a look at the two-sided connection between energy and human well-being. Energy contributes positively to well-being by providing such consumer services as heating, lighting and cooking as well as serving as a necessary input to economic production. But the costs of energy—including not only the money and other resources devoted to obtaining and exploiting it but also the environmental and sociopolitical impacts—detract from well-being.

For most of human history, the dominant concerns about energy have centered on the benefit

side of the energy–well-being equation. Inadequacy of energy resources or (more often) of the technologies and organizations for harvesting, converting and distributing those resources has meant insufficient energy benefits and hence inconvenience, deprivation and constraints on growth. Energy problems in this category remain the principal preoccupation of the least developed countries, where energy for basic human needs is the main issue; they are also an important concern in the intermediate and newly industrializing countries, where the issue is energy for production and growth.

Aside from having too little energy, it is possible to suffer by paying too much for it. The price may be paid in excessive diversion of capital, labor and income from nonenergy needs (thereby producing inflation and reducing living standards), or it may be paid in excessive environmental and sociopolitical impacts. For most of the past 100 years, however, the problems of excessive energy costs have seemed less threatening than the problems of insufficient supply. Between 1890 and 1970 the monetary costs of supplying energy and the prices paid by consumers stayed more or less constant or declined, and the environmental and sociopolitical costs were regarded more as local nuisances or temporary inconveniences than as pervasive and persistent liabilities.

All this changed in the 1970's. The oil-price shocks of 1973–1974 and 1979 doubled and then quadrupled the real price of oil on the world market (see Figure 11.2). In 1973 oil constituted nearly half of the world's annual use of industrial energy forms (oil, natural gas, coal, nuclear energy and hydropower as opposed to the traditional energy forms of fuelwood, crop wastes and dung). Inevitably, the rise in oil prices pulled the prices of the other industrial energy forms upward. The results illustrate vividly the perils of excessive monetary costs of energy: worldwide recession, spiraling debt, a punishing blow to the development prospects of the oil-poor countries of the Southern Hemisphere and the imposition in the industrialized nations of disproportionate economic burdens on the poor.

Figure 11.1 STREET FAIR in New York City brings together large numbers of people from a high-technology segment of the world population of 5.3 billion. If the rest of the world's people used energy at the same rate as citizens of the U.S. do, global energy use in 1990 would be more than four times as large as it is.

The early 1970's also marked a transition . ing to grips with the environmental and sociopo. cal costs of energy. Problems of air and water pollution, many of them associated with energy supply and use, were coming to be recognized as pervasive threats to human health, economic well-being and environmental stability. Consciousness of the sociopolitical costs of energy grew when overdependence on oil from the Middle East created foreign-policy dilemmas and even a chance of war, and when India's detonation of a nuclear bomb in 1974 emphasized that spreading competence in nuclear energy can provide weapons as well as electricity.

The 1970's, then, represented a turning point. After decades of constancy or decline in monetary costs—and of relegation of environmental and sociopolitical costs to secondary status—energy was seen to be getting costlier in all respects. It began to be plausible that excessive energy costs could pose threats on a par with those of insufficient supply. It also became possible to think that expanding some forms of energy supply could create costs exceeding the benefits.

The crucial question at the beginning of the 1990's is whether the trend that began in the 1970's will prove to be temporary or permanent. Is the era of cheap energy really over, or will a combination of new resources, new technologies and changing geopolitics bring it back? One key determinant of the answer is the staggering scale of energy demand brought forth by 100 years of unprecedented population growth, coupled with an equally remarkable growth in per capita demand for industrial energy forms. Supplying energy at rates in the range of 10 terawatts (one terawatt is one billion watts), first achieved in the late 1960's, is an enterprise of enormous scale. The way it was done in 1970 required the harvesting, processing and combustion of some three billion metric tons of coal and lignite, some 17 billion barrels of oil, more than a trillion cubic meters of natural gas and perhaps two billion cubic meters of fuelwood. It entailed the use of dirty coal as well as clean; undersea oil as well as terrestrial; deep gas as well as shallow; mediocre hydroelectric sites as well as good ones; and deforestation as well as sustainable fuelwood harvesting.

The greatest part of the past century's growth in industrial energy forms was supplied by oil and natural gas—the most accessible, versatile, transportable and inexpensive chemical fuels on the planet (see Figure 11.3). The century's cumulative consumption of some 200 terawatt-years of oil and

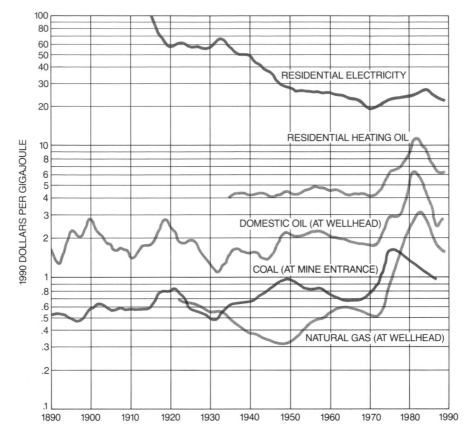

Figure 11.2 ENERGY COSTS in the U.S. over the past century are charted, in constant U.S. dollars. The fuel-price shocks of the 1970's were precipitated by the Organization of Petroleum Exporting Countries (OPEC) but reflect an underlying reality: cheap (easily recoverable) oil and natural gas are already gone in most of the world. Although fuel substitution and conservation brought the price of OPEC oil down in the early 1980's, prices are not likely ever to reach their preshock lows. Electricity was less affected by the price shocks because of the limited role of oil in generating electricity and the modest contribution of fuel costs to the total cost of supplying electricity.

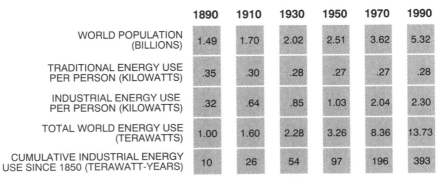

	1890	1910	1930	1950	1970	1990
WORLD POPULATION (BILLIONS)	1.49	1.70	2.02	2.51	3.62	5.32
TRADITIONAL ENERGY USE PER PERSON (KILOWATTS)	.35	.30	.28	.27	.27	.28
INDUSTRIAL ENERGY USE PER PERSON (KILOWATTS)	.32	.64	.85	1.03	2.04	2.30
TOTAL WORLD ENERGY USE (TERAWATTS)	1.00	1.60	2.28	3.26	8.36	13.73
CUMULATIVE INDUSTRIAL ENERGY USE SINCE 1850 (TERAWATT-YEARS)	10	26	54	97	196	393

Figure 11.3 TRENDS in population and energy use per person account for the past century's rapid growth of world energy demand. Industrial energy forms are mainly coal, oil and natural gas, with smaller contributions from hydropower and nuclear energy. Traditional fuels are wood, crop wastes and dung. A terawatt is equal to a billion tons of coal or five billion barrels of oil per year. Data were compiled by the author.

gas represented perhaps 20 percent of the ultimately recoverable portion of the earth's endowment of these fuels. If the cumulative consumption of oil and gas continues to double every 15 to 20 years, as it has done for a century, the initial stock will be 80 percent depleted in another 30 or 40 years.

Except for the huge pool of oil underlying the Middle East, the cheapest oil and gas are already gone. The trends that once held costs at bay against cumulative depletion, that is, new discoveries and economies of scale in processing and transport, have played themselves out. Even if a few more giant oil fields are discovered, they will make little difference against consumption on today's scale. Oil and gas will have to come increasingly, for most countries, from smaller and more dispersed fields, from offshore and Arctic environments, from deeper in the earth and from imports whose reliability and affordability cannot be guaranteed.

There are, as the preceding chapters have shown, a variety of other energy resources that are more abundant than oil and gas. Coal, solar energy and fission and fusion fuels are the most important ones. But they all require elaborate and expensive transformation into electricity or liquid fuels in order to meet society's needs. None has very good prospects for delivering large quantities of fuel at costs comparable to those of oil and gas prior to 1973 or large quantities of electricity at costs comparable to those of the cheap coal-fired and hydropower plants of the 1960's. It appears, then, that expensive energy is a permanent condition, even without allowing for its environmental costs.

The capacity of the environment to absorb the effluents and other impacts of energy technologies is itself a finite resource (see Figure 11.4). The finitude is manifested in two basic types of environmental costs. "External" costs are those imposed by environmental disruptions on society but not reflected in the monetary accounts of the buyers and sellers of the energy. "Internalized" costs are increases in monetary costs imposed by measures, such as pollution-control devices, aimed at reducing the external costs.

Both types of environmental costs have been rising for several reasons. First, the declining quality of fuel deposits and energy-conversion sites to which society must now turn means more material must be moved or processed, bigger facilities must be constructed and longer distances must be traversed. Second, the growing magnitude of effluents from energy systems has led to saturation of the ment's capacity to absorb such effluents wit disruption.

Third, the monetary costs of controlling pollution tend to increase with the percentage of pollutant removed. The combination of higher energy-use rates, lower resource quality and an already stressed environment requires that an increasing percentage be removed just to hold damages constant. Consequently, internalized costs must rise. And fourth, growing public and political concern for the environment has lengthened the time required for siting, building and licensing energy facilities, and has increased the frequency of mid-project changes in design and specifications, forcing costs still further upward.

It is difficult to quantify the total contribution of all these factors to the monetary costs of energy supply, in part because factors not related to the environment are often entwined with environmental ones. For example, construction delays have been caused not just by regulatory constraints but also by problems of engineering, management and quality control. Nevertheless, it seems likely that in the U.S. actual or attempted internalization of environmental impacts has increased the monetary costs of supplying petroleum products by at least 25 percent during the past 20 years and the costs of generating electricity from coal and nuclear power by 40 percent or more.

Despite these expenditures, the remaining uninternalized environmental costs have been substantial and in many cases are growing. Those of greatest concern are the risk of death or disease as a result of emissions or accidents at energy facilities and the impact of energy supplies on the global ecosystem and on international relations.

The impacts of energy technologies on public health and safety are difficult to pin down with much confidence. In the case of air pollution from fossil fuels, in which the dominant threat to public health is thought to be particulates formed from sulfur dioxide emissions, a consensus on the number of deaths caused by exposure has proved impossible. Widely differing estimates result from different assumptions about fuel composition, air-pollution control technology, power-plant siting in relation to population distribution, meteorological conditions affecting sulfate formation and, above all, the relation between sulfate concentrations and disease.

Large uncertainties also apply to the health and safety impacts of nuclear fission. In this case, differ-

AFFECTED QUANTITY	NATURAL BASELINE	HUMAN DISRUPTION INDEX	SHARE OF HUMAN DISRUPTION CAUSED BY:			
			INDUSTRIAL ENERGY	TRADITIONAL ENERGY	AGRICULTURE	MANUFACTURING, OTHER
LEAD FLOW	25,000 TONS/YEAR	15	63% FOSSIL-FUEL BURNING, INCLUDING ADDITIVES	SMALL	SMALL	37% METALS PROCESSING, MANUFACTURING, REFUSE BURNING
OIL FLOW TO OCEANS	500,000 TONS/YEAR	10	60% OIL HARVESTING, PROCESSING, TRANSPORT	SMALL	SMALL	40% DISPOSAL OF OIL WASTES
CADMIUM FLOW	1,000 TONS/YEAR	8	13% FOSSIL-FUEL BURNING	5% BURNING TRADITIONAL FUELS	12% AGRICULTURAL BURNING	70% METALS PROCESSING, MANUFACTURING, REFUSE BURNING
SO$_2$ FLOW	50 MILLION TONS/YEAR	1.4	85% FOSSIL-FUEL BURNING	.5% BURNING TRADITIONAL FUELS	1% AGRICULTURAL BURNING	13% SMELTING, REFUSE BURNING
METHANE STOCK	800 PARTS PER BILLION	1.1	18% FOSSIL-FUEL HARVESTING AND PROCESSING	5% BURNING TRADITIONAL FUELS	65% RICE PADDIES, DOMESTIC ANIMALS, LAND CLEARING	12% LANDFILLS
MERCURY FLOW	25,000 TONS/YEAR	.7	20% FOSSIL-FUEL BURNING	1% BURNING TRADITIONAL FUELS	2% AGRICULTURAL BURNING	77% METALS PROCESSING, MANUFACTURING, REFUSE BURNING
NITROUS OXIDE FLOW	10 MILLION TONS/YEAR	.4	12% FOSSIL-FUEL BURNING	8% BURNING TRADITIONAL FUELS	80% FERTILIZER, LAND CLEARING, AQUIFER DISRUPTION	SMALL
PARTICLE FLOW	500 MILLION TONS/YEAR	.25	35% FOSSIL-FUEL BURNING	10% BURNING TRADITIONAL FUELS	40% AGRICULTURAL BURNING, WHEAT HANDLING	15% SMELTING, NONAGRICULTURAL LAND CLEARING, REFUSE BURNING
CO$_2$ STOCK	280 PARTS PER MILLION	.25	75% FOSSIL-FUEL BURNING	3% NET DEFORESTATION FOR FUELWOOD	15% NET DEFORESTATION FOR LAND CLEARING	7% NET DEFORESTATION FOR LUMBER, CEMENT MANUFACTURING

Figure 11.4 ENERGY SUPPLY accounts for a major share of human impact on the global environment. Most impacts can be characterized as alterations to preindustrial flows or to stocks of environmentally active substances (natural baselines). The human disruption index is the magnitude of the human-generated alteration divided by the baseline. Impacts shown here, except oil flows, all involve flows to or stocks in the atmosphere. The estimates are based on several sources and are approximate.

ing estimates result in part from differences among sites and reactor types, in part from uncertainties about emissions from fuel-cycle steps that are not yet fully operational (especially fuel reprocessing and management of uranium-mill tailings) and in part from different assumptions about the effects of exposure to low-dose radiation. The biggest uncertainties, however, relate to the probabilities and consequences of large accidents at reactors, at reprocessing plants and in the transport of wastes.

Altogether the ranges of estimated hazards to public health from both coal-fired and nuclear-power plants are so wide as to extend from negligible to substantial in comparison with other risks to the population (see Figure 11.5). There is little basis, in these ranges, for preferring one of these energy

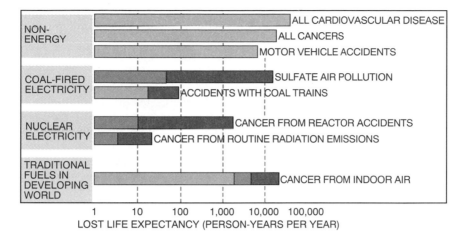

LOST LIFE EXPECTANCY (PERSON-YEARS PER YEAR)

1   10   100   1,000   10,000   100,000

Figure 11.5 RELATIVE MAGNITUDES of some energy and nonenergy risks are shown for a population of one million. Lost life expectancy is measured as the number of deaths per year times the number of years lost per death. Nonenergy data are based on actual U.S. mortality statistics (*blue*). Energy risks are calculated assuming that all electricity use (or household energy use in the developing world) comes from the indicated source. Minimum (*green*) to intermediate (*orange*) to maximum (*red*) estimates of calculated risks are based on various assumptions about sites, dose-response relations and accident probabilities (in the nuclear case).

sources over the other. For both, the very size of the uncertainty is itself a significant liability.

Often neglected, but no less important, is the public health menace from traditional fuels widely used for cooking and water heating in the developing world. Perhaps 80 percent of global exposure to particulate air pollution occurs indoors in developing countries, where the smoke from primitive stoves is heavily laden with carcinogenic benzopyrene and other dangerous hydrocarbons. A disproportionate share of this burden is borne, moreover, by women (who do the cooking) and small children (who are indoors with their mothers).

The ecological threats posed by energy supply are even harder to quantify than the threats to human health and safety from effluents and accidents. Nevertheless, enough is known to suggest they portend even larger damage to human well-being. This damage potential arises from the combination of two circumstances.

First, civilization depends heavily on services provided by ecological and geophysical processes such as building and fertilizing soil, regulating water supply, controlling pests and pathogens and maintaining a tolerable climate; yet it lacks the knowledge and the resources to replace nature's services with technology. Second, human activities are now clearly capable of disrupting globally the processes that provide these services. Energy supply, both industrial and traditional, is responsible for a striking share of the environmental impacts of human activity. The environmental transition of the past 100 years—driven above all by a 20-fold increase in fossil-fuel use and augmented by a tripling in the use of traditional energy forms—has amounted to no less than the emergence of civilization as a global ecological and geochemical force.

Of all environmental problems, the most threatening and in many respects the most intractable is global climate change. Climate governs most of the environmental processes on which the well-being of 5.3 billion people critically depends. And the greenhouse gases most responsible for the danger of rapid climate change come largely from human endeavors too massive, widespread and central to the functioning of our societies to be easily altered: carbon dioxide (CO_2) from deforestation and the combustion of fossil fuels; methane from rice paddies, cattle guts and the exploitation of oil and natural gas; and nitrous oxides from fuel combustion and fertilizer use.

The only other external energy cost that might match the devastating impact of global climate change is the risk of causing or aggravating large-scale military conflict. One such threat is the potential for conflict over access to petroleum resources.

The danger is thought to have declined since the end of the 1970's, but circumstances are easily imagined in which it could reassert itself — particularly given the current resurgence of U.S. dependence on foreign oil. Another threat is the link between nuclear energy and the spread of nuclear weapons. The issue is hardly less complex and controversial than the link between carbon dioxide and climate; many analysts, including me, think it is threatening indeed.

What are the prospects for abating these impacts? Clearly, the choices are to fix the present energy sources or to replace them with others having lower external costs.

As for fixing fossil fuels, it appears that most of their environmental impact (including the hazards of coal mining and most of the emissions responsible for health problems and acid precipitation) could be substantially abated at monetary costs amounting to additions of 30 percent or less to the current U.S. costs of fossil fuels or electricity generated from them. Still, a massive investment in retrofitting or replacing existing facilities and equipment would be needed, representing a particular barrier in parts of the world where capital is scarce and existing facilities and equipment are far below current U.S. standards. The carbon dioxide problem is much harder: replacing coal with natural gas, which releases less CO_2 per gigajoule, is at best a short-term solution, and capturing and sequestering the CO_2 from coal and oil would require revamping much of the world's fuel-burning technology, at huge cost.

Nuclear energy is incomparably less disruptive climatologically and ecologically than fossil fuels are, but its expanded use is unlikely to be accepted unless a new generation of reactors with demonstrably improved safety features is developed, unless radioactive wastes can be shown to be manageable in the real world and not just on paper and unless the proliferation issue is decisively resolved. I believe the first two conditions could be met, at least for nonbreeder reactors, without increasing the already high costs of nuclear electricity by more than another 25 percent. I think the third can be accomplished only by internationalizing a substantial part of the nuclear-energy enterprise, an approach blocked much more by political difficulties than by monetary costs. Fusion can, in principle, reduce the safety, waste and proliferation hazards of fission, but it is not yet clear how soon, by how much and at what monetary cost.

Biomass energy, if replaced continuously by new growth, avoids the problem of net CO_2 production, but the costs of controlling the other environmental impacts of cultivation, harvesting, conversion and combustion of biomass will be substantial. Just bringing the consequence of today's pattern of biomass energy use under control, given its contribution to deforestation and air-pollution problems, will require huge investments of time and money. The tripling or quintupling of biomass supplies foreseen by some would be an even more formidable task, fraught with environmental as well as economic difficulties.

The superabundant long-term option whose external environmental costs are most clearly controllable is direct harnessing of sunlight, but it is now the most expensive of the long-term options and may remain so. The decision to pay the monetary costs of solar energy, if it is made, will represent the ultimate internalization of the environmental costs of the options that solar energy would displace.

There is much reason to think, then, that the energy circumstances of civilization is changing in fundamental rather than superficial ways. The upward trend in energy costs is solidly entrenched, above all because of environmental factors. It is quite plausible, in fact, given existing energy-supply systems, end-use technologies and end-use patterns, that most industrialized nations are near or beyond the point where further energy growth will create greater marginal costs than benefits. "Full speed ahead" is no longer a solution.

Instead we will need transitions in energy-supply systems and patterns of end use just to maintain current levels of well-being; without such transitions, cumulative consumption of high-grade resources and the diminished capacity of the environment to absorb energy's impacts will lead to rising total costs even at constant rates of use. Providing for economic growth without environmental costs that undermine the gains will require even faster transitions to low-impact energy-supply technologies and higher end-use efficiency.

Although the situation poses formidable challenges, it is likely that the most advanced industrialized nations are rich enough and technologically capable enough to master most of the problems. The richest countries could, if they chose, live with low or even negative energy growth by milking increases in economic well-being from efficiency increases, and they could pay considerably higher energy prices to finance the transition to environmentally less disruptive energy-supply technologies. But so far there is little sign of this actually happening. And whether it could be managed in the Soviet Union and Eastern Europe, even in principle, without massive help from the West is more problematic.

Still more difficult is the situation in the less developed countries (LDC's). They would like to industrialize the way the rich did, on cheap energy, but they see the prospects of doing so undermined by high energy costs—whether imposed by the world oil market or by a transition to cleaner energy options. An acute shortage of capital accentuates their tendency to choose options that are cheapest in terms of monetary costs, and they see the local environmental impacts of cheap, dirty energy as a necessary trade for meeting basic human needs (with traditional energy forms) and generating economic growth (with industrial ones).

Although the LDC share of world energy use is modest today, the demographics and economic aspirations of these countries represent a huge potential for energy growth. If this growth materializes and comes mainly from fossil fuels, as most of these countries now anticipate, it will add tremendously to the atmospheric burdens of CO_2 and other pollutants both locally and globally. And while they resent and resist the go-slow approach to energy growth that global environmental worries have fostered in many industrialized nations, the LDC's are, ironically, more vulnerable to global environmental change: they have smaller food reserves, more marginal diets, poorer health and more limited resources of capital and infrastructure with which to adapt.

Global climate change could have profound consequences for the nations of the Southern Hemisphere: more dry-season droughts, more wet-season floods, more famine and disease, perhaps hundreds of millions of environmental refugees. Even if the North suffered less from the direct effects of climate change because of the greater capacities of industrialized societies to adapt, the world is too interconnected by trade, finance, resource interests, politics, porous borders and possibilities for venting frustrations militarily.

How should society respond to the changing and increasingly alarming interaction between energy and human well-being? How can the energy transition on which civilization has embarked, largely unaware, be steered consciously toward a more supportive and sustainable relation among energy, the economy and the environment?

The first requirement is to develop an improved and shared understanding of where we are, where we are headed and where we would like to go. There needs to be an extended public and indeed international debate on the connections between energy and well-being, supported by a greatly expanded research effort to clarify the evolving pattern of energy benefits and costs. Of course, study and debate will take time. Large uncertainties attend many of the important issues, and some of these will take decades to resolve.

Perhaps, with more information, the situation will seem less threatening and difficult than I have suggested; on the other hand, it could be even more threatening and difficult. In any case, we face the dilemma of action versus delay in an uncertain world: if we wait, our knowledge will improve, but the effectiveness of our actions may shrink; damage may become irreversible, dangerous trends more entrenched, our technologies and institutions even harder to steer and reshape.

The solution to the dilemma is a two-pronged strategy consisting of "no regrets" and "insurance policy" elements. No-regrets actions are those that provide leverage against the dangers we fear but are beneficial even if the dangers do not fully materialize. In contrast, insurance-policy actions offer high potential leverage against uncertain dangers in exchange for only modest investment, although some of that investment may later turn out to have been unnecessary.

One essential no-regrets program is to internalize and reduce the environmental and sociopolitical costs of existing energy sources. High priority should be given to abating emissions of sulfur and nitrogen oxides from fossil fuels and emissions of hydrocarbons and particles from fossil fuels and traditional fuels alike. Technologies for controlling these emissions exist and will more than repay their costs by reducing damage to health, property and ecosystems. Another part of the program should be a carbon tax, the revenues from which could be used to develop and finance technologies for reducing fossil-fuel dependence worldwide. More effort is also needed to increase the safety and decrease the weapon-proliferation potential of contemporary (nonbreeder) nuclear reactors; including the development of better reactor designs and placing the most vulnerable fuel-cycle steps under international control.

I ncreasing the efficiency of energy use (another no-regrets approach) is the most effective way of all to abate environmental impacts. Fossil fuels and uranium saved through efficiency generate no emissions and create no fission products or proliferation hazards. (Efficiency, too, can have environmental impacts, but they are usually smaller, or can be made smaller, than those of the energy sources displaced.) Increased efficiency is also the most economical option in monetary terms and the most rapidly expandable, and its ultimate potential is both enormous and sustainable. The main obstacle is educating the vast numbers of individual energy consumers, whose actions hold the key to many of the potential gains, and then providing them with the capital to take advantage of more efficient technologies.

Also crucial to a sensible energy strategy is the acceleration of research (see Figure 11.6), and development on long-term energy alternatives: sunlight, wind, ocean heat, and biomass; the geothermal energy that is ubiquitous in the earth's crust at great depth; fission breeder reactors; fusion; and advanced approaches to energy efficiency. The research should emphasize not only the attainment of economical ways to harness these resources but also the prospects for minimizing their environmental costs. Investing in such research qualifies as an insurance approach in that we do not yet know which of the options will be needed or how soon. Some of the money will be wasted, in the sense that some of the options will never be exploited. But the funding required to develop these alternatives to the point that we can choose intelligently between them is

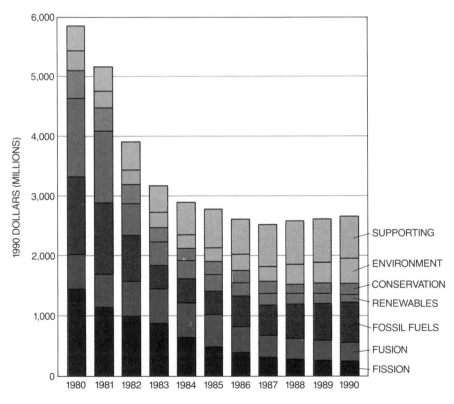

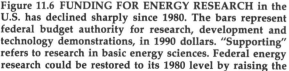

Figure 11.6 FUNDING FOR ENERGY RESEARCH in the U.S. has declined sharply since 1980. The bars represent federal budget authority for research, development and technology demonstrations, in 1990 dollars. "Supporting" refers to research in basic energy sciences. Federal energy research could be restored to its 1980 level by raising the gasoline tax a mere three cents a gallon. Note also that the U.S. military spends more than 100 times the total energy-research budget for "insurance" against events that are far less likely to occur than global changes demanding new energy options.

modest compared with the potential costs of having too few choices.

Building East-West and North-South cooperation on energy and environmental issues, a no-regrets strategy that will help no matter how the future unfolds, might begin with increased cooperation on energy research. Such collaboration could alleviate the worldwide funding squeeze for such research by eliminating needless duplication, sharing diverse specialized strengths and dividing the costs of large projects. (Until now nuclear fusion has been the only area of energy research that has enjoyed major international cooperation.) It is especially important that cooperation in energy research include North-South collaborations on energy technologies designed for application in developing countries.

International cooperation on understanding and controlling the environmental impacts of energy supply is also extremely important, because many of the most threatening problems are precisely those that respect no boundaries. Air and water pollution from Eastern Europe and the Soviet Union reach across Western Europe and into the Arctic, and the environmental impacts of energy supply in China

and India, locally debilitating at today's levels of energy use, could become globally devastating at tomorrow's. But pleas from the rich countries to solve global environmental problems through global energy restraint will fall on deaf ears in the least developed and economically intermediate countries unless the first group can find ways to help the last two achieve increased economic well-being and environmental protection at the same time.

Concerning carbon dioxide, the best hope is that a no-regrets approach to energy efficiency—together with reforestation and afforestation efforts that also fall in the no-regrets category—will be sufficient to stabilize CO_2 emissions even as we wait for the necessarily slower transition to environmentally, economically and politically acceptable non-carbon-based energy sources. But it would be imprudent to assume that no-regrets approaches will suffice. We need more insurance, beyond the research advocated above, to protect us against the possibility that rapid and severe climate change might necessitate an accelerated retreat from fossil

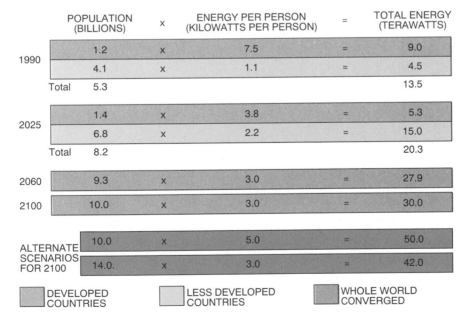

	POPULATION (BILLIONS)	x	ENERGY PER PERSON (KILOWATTS PER PERSON)	=	TOTAL ENERGY (TERAWATTS)
1990	1.2	x	7.5	=	9.0
	4.1	x	1.1	=	4.5
Total	5.3				13.5
2025	1.4	x	3.8	=	5.3
	6.8	x	2.2	=	15.0
Total	8.2				20.3
2060	9.3	x	3.0	=	27.9
2100	10.0	x	3.0	=	30.0
ALTERNATE SCENARIOS FOR 2100	10.0	x	5.0	=	50.0
	14.0	x	3.0	=	42.0

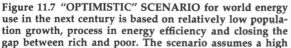 DEVELOPED COUNTRIES   LESS DEVELOPED COUNTRIES   WHOLE WORLD CONVERGED

Figure 11.7 "OPTIMISTIC" SCENARIO for world energy use in the next century is based on relatively low population growth, process in energy efficiency and closing the gap between rich and poor. The scenario assumes a high standard of living can be achieved worldwide at an average rate of energy use of about three kilowatts a person. Nonetheless, energy use in 2060 is double that of 1990.

fuels. We ought to have a contingency plan—carefully researched, cooperatively developed and continuously updated—for reducing global carbon emissions at a rate of 20 percent per decade or more if that proves necessary and if the no-regrets strategies already in place are not adequate.

None of the preceding measures, nor all of them together, will be enough to save us from the folly of failing to stabilize world population (see Figure 11.7). The growth of population aggravates every resource problem, every environmental problem and most social and political problems. Short of catastrophe, world population probably cannot be stabilized at less than nine billion people; without a major effort to limit its growth, the number of human beings on the planet could soar to 14 billion or more.

Supplying 5.3 billion people in 1990 with an average of 2.6 kilowatts per person—a total of 13.7 terawatts—is severely straining the planet's technological, managerial and environmental resources, and crucial human needs are going unmet. Let us suppose, optimistically, that tremendous progress in energy efficiency makes it possible to provide an acceptable standard of living at an average of three kilowatts per person (half the figure for West Germany today). Then nine billion people would use 27 terawatts and 14 billion would use 42; the lower energy-use figure is twice today's, the higher one more than triple today's. Can we expect to achieve even the lower one at tolerable costs? As hard as controlling population growth may be, it is likely to be easier than providing increasing numbers of people with energy (and food and water and much else).

The foregoing prescriptions for taking positive control over the energy transition constitute a demanding and ambitious agenda for national and international action. Little of it will happen unless there is widespread consensus about the nature of the problem, the size of the stakes and the possibilities for action. It is hoped that the chapters in this book will make a contribution toward that end.

Epilogue

*Moving toward greater
energy efficiency*

. . .

Robert Malpas

s populations expand and strive for ever better standards of living, global energy consumption is on the rise. But the ever increasing demand for energy services creates a paradox as daunting as that confronting Adam and Eve. As humans, we need more of the fruit that energy bears, yet we have begun to fear the environmental effects of eating that fruit. We have also begun to worry about the tree's ability to continue bearing fruit or to produce it at an acceptable price. How, then, can we ensure there is sufficient energy to sustain national growth, meet the needs of the poor and protect the global environment?

The solution to our paradox, fortunately, is simple, but it demands that we become engaged in the battle for greater energy efficiency. Is it not more logical to save a barrel of oil by insulating our homes than to waste that valuable resource by letting heat leak through the walls? Is it not senseless to light commercial buildings at night if no one is there? Is not an automobile that travels 24 miles on a gallon of gasoline better than one that travels half that distance on the same amount of fuel? Energy efficiency must be transformed into a significant global force, supported by people and governments everywhere.

How can I promote such a theme when most of my professional career has been spent working for a major oil company? If one thinks efficiency goes hand in hand with a reduction in oil consumption, then my suggestions might indeed be perceived as a major threat to the petroleum industry. But crude oil supplies are finite, and regional shortages are imminent. There are other reasons, too, besides supply to promote efficiency. The shift to updated technologies promises a new range of business opportunities that will force those of us associated with the energy industry to extract and combust fossil fuels more efficiently.

How might the shift to an efficiency-oriented economy come about? The solution, I believe, lies with the engineering profession. After all, the proposition of consuming less to produce more (by conserving resources and reducing waste) is at the heart of all engineering philosophy. For that reason, engineers are the obvious choice to be vanguards of the efficiency movement.

Engineers have let themselves be transformed from innovators into service providers. But our planet is now facing a crisis of unknown proportions, and sitting on the sidelines is no longer acceptable.

Only engineers can harness the extraordinary advances being made by scientists. "Science," aerodynamicist Theodor von Kármán said, "discovers what is; engineers turn this knowledge into things that have never been." And so it is the engineers who are best equipped to know which breakthroughs in science and technology can be brought to production today and which ones might be brought to production tomorrow.

The public also bears genuine responsibility. People, after all, must be willing to embrace new technologies as they are devised. Industrialized nations assume, sometimes erroneously, that technology will come to the rescue on every issue. In a number of cases, certainly, their faith has not been misguided. People believe that technology will extend the finiteness of oil (as indeed it has) and that it will reduce the energy needed to perform a given task — without any action on their part. They also think that environmental concerns will be solved by the cavalry — technology — riding over the hill. Yet it is people who control the market through their willingness to buy or not to buy.

Some success has already been achieved. The apparently incontrovertible one-to-one relation between energy growth and economic growth has been broken. New aircraft operate 20 percent more efficiently than older models do. In Europe, high-speed, 185-mile-per-hour trains are becoming more popular. In the U.S., the concept of energy-efficient homes is gaining ground. Natural gas, a more efficient and cleaner fossil fuel than coal, has an expanding role in power generation around the world. Still far more can be achieved, given the proper incentives to harness existing technology while promoting future technologies.

To begin with, energy efficiency must rise to the top of the global agenda and remain there for at least the next decade. One way to bring the topic to greater prominence might be to call a series of technical meetings that would specifically address issues relevant to energy efficiency. The Council of Academies of Engineering and Technological Sciences, for example (which represents nine countries), meets several times a year to discuss subjects of global importance. Surely it is time for such powerful organizations to speak out in support of efficiency.

Other approaches bear serious consideration. The public should be informed about the long-term penalties and benefits of day-to-day energy decisions. Perhaps a new measurement, comparable to the gross national product (GNP), can be devised to track national energy efficiency (NEE) and heighten awareness of the costs underlying the services that energy provides. We should press to have all energy-consuming equipment rated for efficiency and to have those ratings visibly displayed, whether they are for domestic, commercial or industrial equipment.

Finally, there is an urgent need to promote research in energy conservation. Major centers where such studies are already under way — Harvard and Princeton, the University of California at Berkeley and the World Resources Institute — deserve increased support; new centers in other regions of the world must also be founded and adequately maintained.

All of these efforts will founder if the industrialized nations fail to share critical innovations with the developing countries. Unless the nations in the Southern Hemisphere adopt more efficient technologies at the outset and so avoid the heavy consumption of fossil fuels, they will consume energy at a rate that will obliterate the gains won through efficiency in the Northern Hemisphere.

A simple, powerful equation was imparted to me several years ago by a colleague. "Change," my friend said, results from "dissatisfaction, vision, and practical first steps." Dissatisfaction involves the feeling that we can do better, rather than just complain about how terrible things are. Vision is, of course, what engineers and technologists can provide by articulating what is possible. The practical first step is what engineers must fashion. As one of them, I am willing to accept that responsibility. Will my colleagues join me in taking a step forward?

The Authors

GED R. DAVIS ("Energy for Planet Earth") is head of energy in group planning for Shell International Petroleum Company Limited in London. He has a bachelor's degree in mining engineering from London University and master's degrees in economics from the London School of Economics and in engineering and economics from Stanford University. For the past decade, he has undertaken many environmental analyses for Shell and in particular has examined the ways in which global energy industries might adapt to a sustainable world.

ARNOLD P. FICKETT, CLARK W. GELLINGS and **AMORY B. LOVINS** ("Efficient Use of Electricity") are consultants to the power industry. Fickett is vice-president of the Customer Systems Division at the Electric Power Research Institute (EPRI). He received an M.S. in electrochemistry from Northeastern University. Fickett has more than 30 years of experience in the research, engineering and application of energy-related technologies. Gellings, who is director of the Customer Systems Division at EPRI, has a master's in mechanical engineering from the New Jersey Institute of Technology and a master's in management science from the Stevens Institute of Technology. He spent more than 20 years in energy-related technologies as well as in marketing, forecasting, demand-side management, least-cost planning and conservation. Lovins directs research at Rocky Mountain Institute, a nonprofit resource policy center. He and his wife, L. Hunter Lovins, founded the center in 1982. Winner of the Onassis Foundation's first Delphi prize, he was educated at Harvard and Oxford universities.

RICK BEVINGTON and **ARTHUR H. ROSENFELD** ("Energy for Buildings and Homes") are both buildings experts. Bevington is national marketing and sales manager for Johnson Yokogawa, a U.S.-Japanese venture. Before that he was manager of sales support in the national buildings services division of Johnson Controls in Milwaukee. Rosenfeld is professor of physics at the University of California, Berkeley, and also director of the Center for Building Science at Lawrence Berkeley Laboratory. In 1986 Rosenfeld received the Leo Szilard prize for physics in the public interest. He currently serves on a National Academy of Sciences Panel on the Policy Implications of Greenhouse Warming and advises the California legislature's Committee on Energy Regulation and the Environment.

MARC H. ROSS and **DANIEL STEINMEYER** ("Energy for Industry") have taken a strong interest in industrial energy issues for more than 15 years. Ross is professor of physics at the University of Michigan at Ann Arbor and senior scientist at Argonne National Laboratory. He has a Ph.D. in physics from the University of Wisconsin-Madison and spend 20 years on the theory of fundamental particles before turning to energy and environmental problems. Steinmeyer, an engineering fellow at the Monsanto Company, has master's degrees in economics and engineering from Washington University and Ann Arbor, respectively. He has held a number of process-engineering posts at Monsanto, including that of director of engineering in Brazil and manager of process energy technology. His current work focuses on preventing waste by modifying industrial processes.

DEBORAH L. BLEVISS and **PETER WALZER** ("Energy for Motor Vehicles") study the technology and policy of automotive energy use. Bleviss is executive director of the International Institute for Energy Conservation in Washington, D.C., and author of *The New Oil Crisis and Fuel Economy Technologies: Preparing the Light Transportation Industry for the 1990's* (1988). She received a B.S. in physics from the University of California, Los Angeles, and did additional graduate work at Princeton University. Walzer is head of corporate research at Volkswagen AG and a lecturer at the Technical University of Aachen. He studied economics and aeronautical engineering as an undergraduate and received a doctorate in engineering from the University of Aachen in 1970. After 10 years of research experience, he joined Volkswagen, becoming first director of research on alternative power plants and then executive director of research on engines, electronics, materials and aerodynamic design.

AMULYA K. N. REDDY and **JOSÉ GOLDEMBERG** ("Energy for the Developing World") have frequently published on the subjects of energy, technology and development and have collaborated on several projects. Reddy is chairman of the department of management studies at the Indian Institute of Science in Bangalore and vice-chairman of the Karnataka State Council for Science and Technology. He has also coauthored a textbook on electrochemistry. Goldemberg, the secretary for science and technology of Brazil, has been professor of physics at the University of São Paulo since 1951, where he also served as rector between 1986 and 1989. He is former president of the Energy Company of the state of São Paulo.

WILLIAM U. CHANDLER, ALEXEI A. MAKAROV and **ZHOU DADI** ("Energy for the Soviet Union, Eastern Europe and China") study the production and consumption of energy. Chandler, who holds a B.S. from the University of Tennessee and an M.P.A. from the Kennedy School of Government at Harvard University, is director of the Advanced International Studies Unit at Batelle, Pacific Northwest Laboratories. Makarov has a Ph.D. in energy economics from the Leningrad Polytechnic Institute. He is director of the Energy Research Institute of the Academy of Sciences and State Council on Science and Technology of the Soviet Union. Zhou is director of Energy Systems Analysis of the Energy Research Institute of China. An adviser to the World Bank, he holds a master's degree in engineering from Tsinghua University in Beijing.

WILLIAM FULKERSON, RODDIE R. JUDKINS and **MANOJ K. SANGHVI** ("Energy from Fossil Fuels") are, respectively, associate director for advanced energy systems at the Oak Ridge National Laboratory, manager of the fossil energy program at Oak Ridge and director of industry analysis and forecasts at the Amoco Corporation. Fulkerson oversees several divisions and programs at Oak Ridge, including the one headed by Judkins, who directs the laboratory's research and development relating to fossil energy. Sanghvi counts among his responsibilities the analysis, interpretation and forecasting of the worldwide supply, demand and price of oil, gas and other sources of energy.

WOLF HÄFELE ("Energy from Nuclear Power") is former director general of the Jülich Research Center in West Germany. He was deputy director of the International Institute for Applied Systems Analysis in Laxenburg, Austria, and has served as head of a nuclear-safeguard project and as a scientific adviser to the West German government. He studied technical and theoretical physics in Munich, completed his thesis at the Max Planck Institute for Physics and received his Ph.D. in theoretical physics from the University of Göttingen in 1955. In 1972 Häfele was named a *chevalier de l'ordre des Palmes Academiques*.

CARL J. WEINBERG and **ROBERT H. WILLIAMS** ("Energy from the Sun") share the goal of creating a sustainable energy economy. Weinberg is manager of research and development at the Pacific Gas & Electric Company, where he is involved with wind, solar-thermal, photovoltaic and geothermal systems as well as advanced energy-delivery technologies. He sits on the board of directors of the American Wind Energy Association and the Solar Energy Industries Association. Williams, senior research physicist at the Center for Energy and Environmental Studies at Princeton University, performs research in energy technology assessment and energy policy. He has coauthored seven books, edited four books and written numerous articles on energy. He is a founding board member and past chairman of the American Council for an Energy Efficient Economy.

JOHN P. HOLDREN ("Energy in Transition") is professor of energy and resources at the University of California, Berkeley, and acting chair of the interdisciplinary graduate degree program in energy and resources, which treats problems of energy, resources, development and security in terms of their technological, environmental, economic and sociopolitical components. He received bachelor's and master's degrees in aeronautics and astronautics from the Massachusetts Institute of Technology and his Ph.D. in plasma physics from Stanford University. He is also chair of the executive committee of the Pugwash Conferences on Science and World Affairs. In 1981 Holdren received a MacArthur Foundation prize.

ROBERT MALPAS ("Moving Toward Greater Energy Efficiency") is chairman of PowerGen, a privatized utility that until 1989 was controlled by the Central Electricity Generating Board of England. He was previously managing director of the British Petroleum Company.

Bibliography

1. Energy for Planet Earth

Ausubel, Jesse, and Hedy E. Sladovich, eds. 1989. *Technology and environment*. National Academy Press.

Organisation for Economic Co-operation and Development and International Energy Agency. 1989. *Expert seminar on energy technologies for reducing emissions of greenhouse gases* (collected papers). Paris, April 12–14.

World Energy Conference. 1989. *Energy for tomorrow* (collected papers of the Fourteenth Congress). Montreal, September 17–21.

2. Efficient Use of Electricity

Electric Power Research Institute. 1984–1988. *Demand-side management*, volumes 1–5, EA/EM-3597.

Lovings, Amory B., et al. 1988–1990. *The state of the art: Lighting*. Implementation Papers, Rocky Mountain Institute's COMPETITEK Service.

National Association of Regulatory Commissioners. 1988. *Least-cost utility planning handbook*.

Lovings, Amory B., et al. 1989. *The state of the art: Drivepower*. Implementation Papers, Rocky Mountain Institute's COMPETITEK Service.

Congress of the World Energy Conference. 1989. *End-use/least-cost investment strategies*. Report no. 2.3.1, September 17–22. Rocky Mountain Institute.

Johansson, T. B., B. Bodlund and R. H. Williams, eds. 1989. *Electricity*. American Council for an Energy Efficient Economy.

Moskovitz, David. 1989. *Profits and progress through least-cost planning*. National Association of Regulatory Commissioners.

Electric Power Research Institute. 1990. *Efficient electricity use: Estimates of maximum energy savings*. CU-6746 (March).

3. Energy for Buildings and Homes

Rosenfeld, Arthur H., and David Hafemeister. 1988. Energy-efficient buildings. *Scientific American* 258 (April): 56–63.

Chandler, William U., Howard Geller and Marc R. Ledbetter. 1988. *Energy efficiency: A new agenda*. American Council for an Energy Efficient Economy.

Braun, J. E. 1990. Reducing energy costs and peak electrical demand through optimal control of building thermal storage. *ASHRAE Transactions* 96.

Rosenfeld, Arthur. 199 . Energy-efficient buildings in a warming world. In *Energy and the environment in the 21st century*, ed. Jefferson Tester. MIT Press.

Koomey, Jonathan, and Arthur Rosenfeld. 199 . Revenue-neutral incentives for efficiency and environmental quality. *Contemporary Policy Issues*.

4. Energy for Industry

Haustein, H., and E. Neuwirth. 1982. Long waves in world industrial production, energy consumption, innovations, inventions, and patents, and their identification by spectral analysis. *Technological Forecasting and Social Change* 22:53–89.

Steinmeyer, D. E. 1982. Take your pick: Capital or energy. *Chemtech* 12 (March): 188–192.

Kenney, W. F. 1984. *Energy conservation in the process industry*. Academic Press.

Steinmeyer, D. E. 1984. Process energy conservation. In *Encyclopedia of chemical technology*, supplement volume, ed. Kirk Othmer. John Wiley & Sons, Inc.

Larson, Eric D., Marc H. Ross and Robert H. Williams. 1986. Beyond the era of materials. *Scientific American* 254 (June): 34–41.

5. Energy for Motor Vehicles

Gray, Charles L., Jr., and Frank von Hippel. 1981. The fuel economy of light vehicles. *Scientific American* 244 (May): 48–59.

1984. *Proceedings of the Vecon '84 conference on fuel efficient power trains and vehicles*. Institute of Mechanical Engineers.

Sperling, Daniel, ed. 1989. *Alternative transportation fuels: An environmental and energy solution*. Quorum Books.

Seiffert, U., and Peter Walzer. 1990. *Automotive technology of the future*. SAE.

Libertore, Robert G. 1990. An industry view: Market incentives. *Forum for Applied Research and Public Policy* 5 (Spring): 19–22.

Greene, David L. 1990. Technology and fuel efficiency. *Forum for Applied Research and Public Policy* 5 (Spring): 23–29.

6. Energy for the Developing World

Larson, Eric D., Marc H. Ross and Robert H. Williams. 1986. Beyond the era of materials. *Scientific American* 254 (June): 34–41.

Goldemberg, José, Thomas B. Johansson, Amulya K. N. Reddy and Robert H. Williams. 1988. *Energy for a sustainable world*. Wiley Eastern Limited.

7. Energy for the Soviet Union, Eastern Europe and China

Kornai, János. 1986. *Contradictions and dilemmas: Studies in the socialist economy and society*. MIT Press.

1989. *Energy in China 1989*. Ministry of Energy, Beijing.

Makarov, A. A., and I. A. Bashmakov. 1990. *The Soviet Union: A strategy of energy development with minimum emission of greenhouse gases*. Pacific Northwest Laboratories.

Sitnicki, S., A. Szpilewicz, J. Michna, J. Juda and K. Budzinski. 1990. *Poland: Opportunities for carbon emission control*. Pacific Northwest Laboratories.

Chandler, W. U., ed. 199 . *Carbon emissions control strategies: Case studies in international cooperation*. Conservation Foundation.

8. Energy from Fossil Fuels

Fulkerson, William, et al. 1989. Energy technology R&D: What could make a difference? *Synthesis Report*, part I. ORNL-6541/VI. Oak Ridge National Laboratory.

Smith, Joel B., and Dennis Tirpak, eds. 1989. *The potential effects of global climate change on the United States*. EPA-230-05-89-050, Report to Congress (December).

U.S. Department of Energy. 1990. *National energy strategy, interim report*. DOE/s-0066P (April).

Hendriks, Chris A., Kornelis Blok and Wim C. Turkenburg. 199 . Technology and cost of recovery and storage of carbon dioxide from an integrated gasifier combined cycle plant. *Applied Energy*.

9. Energy from Nuclear Power

Marchetti, Cesare. 1975. Geoengineering and the energy island. In *Second status report of the IIASA Project on energy systems*, eds. W. Häfele et al. International Institute for Applied Systems Analysis RR-76-1.

International Institute for Applied Systems Analysis. 1981. *Energy in a finite world*, volumes 1 and 2. Ballinger Publishing Company.

Weinberg, Alvin M. 1985. Nuclear energy and proliferation: A longer perspective. In *The nuclear connection: A reassessment of nuclear power and nuclear proliferation*, eds. Alvin Weinberg et al. Washington Institute Press.

Holdren, John, et al. Exploring the competitive potential of magnetic fusion energy: The interaction of economics with safety and environmental characteristics. *Fusion Technology* 13 (January): 7–56.

Häfele, W. 1989. Technical safety measures and rules in the nuclear field. *Atomwirtschaft, Atomtechnik* (November):

10. Energy from the Sun

Department of Research and Development, Pacific Gas & Electric Company. 1989. *Wind energy group technical papers, 1982–1989*. Report 007.1-89.2 (November 28).

Ogden, Joan M., and Robert H. Williams. 1989. *Solar hydrogen: Moving beyond fossil fuels*. World Resources Institute.

Solar Energy Research Institute. 1990. *The potential of renewable energy*. Interlaboratory White Paper, SERI/TP-260-3674 (March).

Zweibel, Ken. 1990. *Harnessing solar power: The PV challenge*. Plenum Press.

11. Energy in Transition

Smith, Kirk R. 1987. *Biofuels, air pollution, and health: A global review*. Plenum Press.

Holdren, John P. 1987. Global environmental issues related to energy supply. *Energy* 12:975–992.

Lashof, Daniel A., and Dennis A. Tirpak, eds. 1989. *Policy options for stabilizing global climates*. U.S. Environmental Protection Agency.

Sources of the Photographs

INDEX

Page numbers in *italics* indicate illustrations.